NATIONAL ACADEMIES
Sciences
Engineering
Medicine

NATIONAL ACADEMIES PRESS
Washington, DC

The Role of Net Metering in the Evolving Electricity System

Committee on the Role of Net Metering in the Evolving Electricity System

Board on Energy and Environmental Systems

Division on Engineering and Physical Sciences

Board on Environmental Change and Society

Division of Behavioral and Social Sciences and Education

Consensus Study Report

NATIONAL ACADEMIES PRESS 500 Fifth Street, NW Washington, DC 20001

This study was supported by Contract AWD-001207 with the Department of Energy. Any opinions, findings, conclusions, or recommendations expressed in this publication do not necessarily reflect the views of any organization or agency that provided support for the project.

International Standard Book Number-13: 978-0-309-69331-8
International Standard Book Number-10: 0-309-69331-4
Digital Object Identifier: https://doi.org/10.17226/26704

This publication is available from the National Academies Press, 500 Fifth Street, NW, Keck 360, Washington, DC 20001; (800) 624-6242 or (202) 334-3313; http://www.nap.edu.

Printed in the United States of America.

Suggested citation: National Academies of Sciences, Engineering, and Medicine. 2023. *The Role of Net Metering in the Evolving Electricity System*. Washington, DC: The National Academies Press. https://doi.org/10.17226/26704.

The **National Academy of Sciences** was established in 1863 by an Act of Congress, signed by President Lincoln, as a private, nongovernmental institution to advise the nation on issues related to science and technology. Members are elected by their peers for outstanding contributions to research. Dr. Marcia McNutt is president.

The **National Academy of Engineering** was established in 1964 under the charter of the National Academy of Sciences to bring the practices of engineering to advising the nation. Members are elected by their peers for extraordinary contributions to engineering. Dr. John L. Anderson is president.

The **National Academy of Medicine** (formerly the Institute of Medicine) was established in 1970 under the charter of the National Academy of Sciences to advise the nation on medical and health issues. Members are elected by their peers for distinguished contributions to medicine and health. Dr. Victor J. Dzau is president.

The three Academies work together as the **National Academies of Sciences, Engineering, and Medicine** to provide independent, objective analysis and advice to the nation and conduct other activities to solve complex problems and inform public policy decisions. The National Academies also encourage education and research, recognize outstanding contributions to knowledge, and increase public understanding in matters of science, engineering, and medicine.

Learn more about the National Academies of Sciences, Engineering, and Medicine at **www.nationalacademies.org**.

Consensus Study Reports published by the National Academies of Sciences, Engineering, and Medicine document the evidence-based consensus on the study's statement of task by an authoring committee of experts. Reports typically include findings, conclusions, and recommendations based on information gathered by the committee and the committee's deliberations. Each report has been subjected to a rigorous and independent peer-review process and it represents the position of the National Academies on the statement of task.

Proceedings published by the National Academies of Sciences, Engineering, and Medicine chronicle the presentations and discussions at a workshop, symposium, or other event convened by the National Academies. The statements and opinions contained in proceedings are those of the participants and are not endorsed by other participants, the planning committee, or the National Academies.

Rapid Expert Consultations published by the National Academies of Sciences, Engineering, and Medicine are authored by subject-matter experts on narrowly focused topics that can be supported by a body of evidence. The discussions contained in rapid expert consultations are considered those of the authors and do not contain policy recommendations. Rapid expert consultations are reviewed by the institution before release.

For information about other products and activities of the National Academies, please visit www.nationalacademies.org/about/whatwedo.

COMMITTEE ON THE ROLE OF NET METERING IN THE EVOLVING ELECTRICITY SYSTEM

JANET GAIL BESSER, Independent Expert, *Chair*
ANURADHA M. ANNASWAMY, Massachusetts Institute of Technology
GALEN BARBOSE, Lawrence Berkeley National Laboratory
SEVERIN BORENSTEIN, University of California, Berkeley[1]
MARILYN A. BROWN, NAE/NAS, Georgia Institute of Technology
MELICIA V. CHARLES, Mainspring Energy[2]
MOHIT CHHABRA, Natural Resources Defense Council
ELENA M. KRIEGER, PSE Healthy Energy
JOSHUA M. PEARCE, Western University
AUTUMN F. PROUDLOVE, NC Clean Energy Technology Center
VARUN RAI, The University of Texas at Austin
MOHAMMAD SHAHIDEHPOUR, NAE, Illinois Institute of Technology
NICOLE D. SINTOV, The Ohio State University
THOMAS S. STANTON, National Regulatory Research Institute (Retired)
TERRENCE G. SURLES, Independent Consultant
SUSAN F. TIERNEY, Analysis Group

Staff

K. JOHN HOLMES, Board Director and Scholar, Board on Energy and Environmental Systems (BEES)
BRENT HEARD, Study Director and Program Officer, BEES
DANIEL TALMAGE, Program Officer, Board on Environmental Change and Society
REBECCA DEBOER, Research Associate, BEES
ELI NASS, Research Assistant, DEPS
JASMINE BRYANT, Research Assistant, BEES
KAIA RUSSELL, Program Assistant, BEES

[1] Resigned from the committee August 2022.
[2] Resigned from the committee April 2023.

Reviewers

This Consensus Study Report was reviewed in draft form by individuals chosen for their diverse perspectives and technical expertise. The purpose of this independent review is to provide candid and critical comments that will assist the National Academies of Sciences, Engineering, and Medicine in making each published report as sound as possible and to ensure that it meets the institutional standards for objectivity, evidence, and responsiveness to the study charge. The review comments and draft manuscript remain confidential to protect the integrity of the deliberative process.

We thank the following individuals for their review of this report:

Kate Anderson, National Renewable Energy Laboratory
Paul Centolella, Paul Centolella & Associates, LLC
Peter Fox-Penner, Boston University
Matthew Freedman, The Utility Reform Network
Kenneth Gillingham, Yale University
Lynne Kiesling, University of Colorado Denver
Sarah Kurtz, NAE, University of California, Merced
Mark LeBel, Regulatory Assistance Project
Suzanne Leta, SunPower
Amparo Nieto, PA Consulting
Karl Rábago, Rábago Energy LLC
Timothy D. Wilson, NAS, University of Virginia

Although the reviewers listed above have provided many constructive comments and suggestions, they were not asked to endorse the conclusions or recommendations of this report, nor did they see the final draft before its release. The review of this report was overseen by Granger Morgan, NAS, Carnegie Mellon University, and Anjan Bose, NAE, Washington State University. They were responsible for making certain that an independent examination of this report was carried out in accordance with the standards of the National Academies and that all review comments were carefully considered. Responsibility for the final content of the report rests entirely with the authoring committee and the National Academies.

Contents

Preface

In its 2020 appropriations for the Department of Energy (DOE), the U.S. Congress directed the National Academies of Science, Engineering, and Medicine to appoint an ad hoc committee of experts to study the issues associated with net metering, including the medium- to long-term impacts of net metering on the electricity grid and consumers. This report responds to that direction. It is intended to provide useful guidance to policymakers, legislators, regulators, utilities and their governing boards, distributed generation providers, electricity customers, and other stakeholders interested in and affected by how the electricity system can continue to provide safe, affordable, and reliable service as it evolves to become increasingly decarbonized, equitable, and resilient.

I wish to express my appreciation to the members of the committee for their dedicated and thoughtful contributions to the study and to its timely preparation. This committee's careful and nuanced discussions informed the perspective this report takes towards net metering.

The committee is grateful to DOE as well as the representatives of the various organizations that provided presentations, comments, and active participation throughout the study: Michele Boyd (DOE), Scott Burger (Form Energy), Stephen Campbell (GRID Alternatives), Gabriel Chan (University of Minnesota), Colton Ching (Hawaiian Electric), Sachu Constantine (Vote Solar), Jon Creyts (Rocky Mountain Institute [RMI]), Thad Culley (Sunrun), Gavin Dillingham (Houston Advanced Research Center), Mark Dyson (RMI), Matthew Freedman (The Utility Reform Network), Sean Gallagher (Solar Energy Industries Association), Marissa Gillett (Connecticut's Public Utilities Regulatory Authority), Rachel Gold (RMI), Allison Hamilton (NRECA), Anne Hoskins (Sunrun), Lon Huber (Duke Energy Solutions), Christopher Irwin (DOE), Mike Judge (Coalition for Community Solar Access), Mark LeBel (Regulatory Assistance Project), Roger Lin (Center for Biological Diversity), Shelby Linton-Keddie (Edison Electric Institute), Kevin Lynn (DOE), Michelle Moore (Groundswell), Lee Peterson (CohnReznick), Karl Rábago (Rábago Energy LLC), Daniel Scripps (Michigan Public Service Commission), Benjamin Sovacool (Boston University), Reuven Sussman (American Council for an Energy-Efficient Economy), and Kim Wolske (The University of Chicago).

The committee would also like to thank the National Academies staff—John Holmes, Brent Heard, Daniel Talmage, Rebecca DeBoer, Jasmine Bryant, Kaia Russell, and Eli

Nass—for their work to support this study. Finally, we thank Cathy Gruber and the executive office reports staff of the Division on Engineering and Physical Sciences, who provided valuable help with editing the report, and Katiria Ortiz, who managed the report review process. We hope this report can provide instructive and actionable guidance to policymakers, regulators, and stakeholders in the electricity system as they navigate the evolution of net metering to support a decarbonizing, equitable, and resilient electricity system.

Janet Gail Besser, *Chair*
Committee on the Role of Net Metering
in the Evolving Electricity System

Glossary

This report uses several key terms, which are defined as follows to facilitate a common understanding and foundation for the committee's consideration of net metering and the evolving electricity system.

Avoided cost (AC): Cost that an entity (e.g., a utility) would otherwise incur to produce or acquire an additional unit (kWh) of generation or a substitute (e.g., energy efficiency). Several different methodologies exist for calculating avoided costs. In terms of net metering compensation, avoided cost-based rates are typically the lowest because they include only generation costs. These rates often do not include all of the avoided costs, normally comprising only the generation cost.

Behind the meter (BTM): With respect to net metering, BTM refers to energy devices on the customer's side of the meter which have the potential to export generation and/or discharge storage to the electricity grid. (BTM devices can also include efficiency and demand response technologies that reduce customer usage.)

Clean energy standard (CES): A portfolio standard which mandates that retail sales of electricity to end-use customers contain a given share of electricity sourced from eligible zero- or low-emissions sources.

Distributed energy resources (DER): Energy devices (including generation and storage, as well as energy efficiency and demand response), which are not centrally located or connected into the electricity system, but may be connected at the distribution-level.

Distributed generation (DG): Electricity generated from sources of supply located near the point of use instead of centralized generation sources from power plants. DG facilities are typically interconnected to the distribution system.

Distribution: In the context of this report, this refers to the low-voltage, local electric system.

Economic efficiency: An allocation of resources within a market such that overall value to all participants is maximized while minimizing cost. In order for economic efficiency calculations to be valid, all costs and values must be fully quantified so that the social marginal cost is optimized.

Energy affordability: The affordability of energy bills for all customers accounting for the value of services delivered (see Electric Power Research Institute's 2021 list of Sustainability Priorities).

Energy burden (also, energy cost burden): Refers to the share of a household's income that is spent on energy utilities.

Energy efficiency: A measure or device or process improvement that enables a consumer to achieve his/her needs (e.g., for cooling, space heating, lighting) with less energy. The energy service divided by the energy consumed.

Energy insecurity: The uncertainty that a household might face in being able to make utility payments and the risk of shutoffs due to an inability to pay bills.

Energy poverty: Living in a home that does not have access to enough energy to meet essential needs.

Equity: Equity encompasses:
Distributional equity, encompassing the equal allocation of rights, resources, and/or information.
Procedural equity, focusing on who is represented and engaged in decision-making processes.
Intergenerational equity, considering obligations to future generations.
Structural equity, recognizing the roles and legacies of historical, cultural, and institutional power dynamics and structures.

Fixed charge: An element of an electricity bill that does not vary with the customer's usage.

Grid defection: In the context of this report, this concept includes (a) an electricity customer's actions (e.g., installation of a generation and storage system) that enable the customer to fully disconnect and operate independently from the broader electricity system; and (b) an electricity customer's adoption of on-site generation that serves a major share of that customer's electricity needs. See also Load defection.

Independent system operator (ISO): An entity independent of the owners of generation, transmission, and distribution assets and other market participants that administers the system's reliability and energy-market functions. Similar to a Regional Transmission Organization.

Investor-owned utility (IOU): A utility that is a private company, with shareholders that own the assets. Almost universally, the rates and investments of investor-owned utilities are regulated by state public utility commissions and the Federal Energy Regulatory Commission.

Load defection: Any action that reduces end-user demand (load) (includes self-supply of electricity from generation, energy efficiency, demand response and/or load-shifting, energy conservation, and disconnection from the grid by the customer).

Location based: Rate varies by location on the distribution system or may be a flat rate available to customers in certain zones of the distribution system; typically, an adder or a component of a value-based rate.

Organized markets (or organized wholesale markets): Electricity markets administered by regional transmission organizations or independent system operators.

Public Utilities Commission (PUC): The state agency (sometimes called the Public Service Commission [PSC] or similar) with responsibility to regulate the prices and other terms and conditions under which an electric utility (in particular, an investor-owned utility) may sell electricity to its retail consumers.

Public Utility Regulatory Policies Act (PURPA): A federal statute P.L. 95-617, 92 Stat. 3117, first enacted in 1978 with the purpose of promoting the efficient production of electricity and the use of renewable energy to produce power. PURPA requires utilities to purchase power from Qualifying Facilities at the utility's avoided cost.

Publicly owned utility: In this context of this report, this includes utilities that are cooperatives, municipal utilities, federal and state public utility authorities, special purpose utility districts, and other utilities not owned by private investors.

Qualifying facility (QF): A generating unit with size, fuel, and other attributes that qualifies it to sell its output at the utility's avoided cost as set forth in the Public Utility Regulatory Policies Act and in the associated regulations of the Federal Energy Regulatory Commission.[1]

[1] See https://www.ferc.gov/qf.

Regional transmission organization (RTO): An entity independent of the owners of generation, transmission, and distribution assets and other market participants that administers the reliability, transmission, and energy-market functions of the electricity system. Similar to an independent system operator.

Renewable portfolio standard (RPS): A regulatory mandate requiring that sales of electricity to retail electricity consumers contain a given percentage of electricity production be sourced from eligible renewable generation.

Retail rate: Rates charged for electricity service to the end-use customer. Includes generation, transmission, distribution rates (which may include energy, demand, and/or fixed charges). May include seasonal or other time-of-use variation. The effective retail rate includes all of these costs divided by the amount of energy consumed to be measured in $/kWh.

Storage (electricity storage, energy storage, or stored generation): Energy storage is the capture of energy produced at one time for use at a later time. Electricity storage devices (e.g., batteries, pumped hydroelectric facilities, flywheels, thermal storage equipment) can help to reduce imbalances between energy demand or consumption and energy production.[2]

Time-varying rates (TVR): Electricity rates that vary based on day of week, and time of day, which are more closely tied to the utility's actual cost. May also include seasonal variation.

Transmission: In the context of this report, this refers to the delivery of electricity over the high-voltage electricity system.

Value-based rates: Rates based on the value of solar or other distributed generation. Many different methodologies exist for calculating value-based rates. These rates may include societal and/or non-energy benefits or values.

[2] See https://www.epa.gov/energy/electricity-storage.

Variable charge: An element of a customer's electricity bill based on the amount used for energy (expressed in \$/kWh), or for demand (expressed in \$/kW of peak demand, or both) over the billing period.

Wholesale rate: Rates charged to customers for electricity purchased in wholesale markets for their own use or for resale to retail customers.

Summary

THE ROLE OF NET METERING IN THE EVOLVING ELECTRICITY SYSTEM

Net metering is one element in a broad constellation of factors affecting the electricity system, utility business models, and regulatory frameworks. The key drivers of the evolution of the electricity system include:

- Changing economics that involve both declining costs for current and new customer technologies as well as increasing needs for investment in the grid to integrate them and capture their value for the system as a whole.
- Technology innovations that influence the economics of distributed generation (DG), demand management, storage, information and communications, and power electronics, occurring both on the customer and utility sides of the meter.
- Expanding policy objectives that require utilities and electricity system operators, as well as the economy as a whole, to reduce carbon emissions, improve public health outcomes, address equity, and enhance resilience in the face of more frequent extreme weather events and other electricity system disturbances.
- Increasing customer expectations for new electricity-related products and services to help them manage their energy use and bills, as well as meet their own decarbonization goals.

These drivers are also spurring the evolution of new utility business models, along with the growth in non-utility product and service providers, including DG installers, energy service companies, competitive energy suppliers, microgrid operators, and community choice aggregators, among others, who can be utility partners, complements, or competitors. Succeeding in this increasingly complex environment will require not only new utility business approaches but also new product and service provider approaches, as well as innovation in the regulatory framework within which utilities operate.

In this context, net metering policy has been driving the deployment of behind-the-meter DG to achieve a variety of policy objectives—clean energy, resource diversity, carbon reduction, economic development, equity, and resilience—as well as customer expectations for energy choices and savings. As a result, it is receiving increasing attention and scrutiny as greater quantities of DG and distributed energy resources (DER) are connected to the electricity grid.

Net metering is a means to compensate customers for the output of their behind-the-meter (BTM) DG (most-notably, rooftop solar), some of which they may consume themselves and some of which flows back to the electricity grid. Where implemented, net metering has been successful in supporting DG adoption, particularly solar, leading to BTM DG levels that exceed 10 percent in a handful of states and are forecasted to increase in others.[1] The increasing adoption of distributed solar is one of the factors contributing to rapid and dramatic changes in the electricity system. Net metering needs to evolve because the electricity system is evolving.

Specifically, net metering is a billing mechanism based on retail electricity rates. Ideally, retail rates would reflect the underlying societal costs of electricity.[2] However, many important considerations (such as fairness and simplicity) affect the structure and level of retail rates, which can cause them to deviate from costs. If retail rates cannot be readily adjusted to reflect costs, then the net metering mechanism itself can be changed to compensate BTM DG production based on its value, rather than the retail rate.

As policymakers consider changes to net metering policy to align with the evolving electricity system, it is essential to address equity as these changes present an opportunity to improve outcomes for all electricity system customers. To capture the full value of BTM DG for all customers, investments must also be made to upgrade the grid. Therefore, it will be critical for regulators to establish a framework that allows appropriate cost-recovery mechanisms for these investments. As policymakers further pursue decarbonization, resilience, equity, and other targeted objectives, they should consider looking outside of electricity rates to other sources of funding (e.g., provisions in federal legislation) to target support for solar and DG that can help to accomplish them.

The U.S. Congress directed the National Academies of Sciences, Engineering, and Medicine to appoint an ad hoc committee of experts to study the issues associated with net metering, including the medium- to long-term impacts on the electricity grid and customers.[3] This report is the result of our research and analysis. The committee concludes that the evolution of net metering build on three foundational pillars:

[1] Calculated from statewide residential PV system counts (Wood Mackenzie 2022) and from the 215 total number of residential electricity customers in each state (EIA 2022c), see Figure 3-6 for more detail.

[2] In economics, terms, when prices are based on costs, customers can make efficient decisions about their electricity consumption and production.

[3] The full statement of task can be seen in Chapter 1.

- Recognize that the future electricity system is and will become even more interdependent among its constituent parts, including DG at the grid edge. Therefore, net metering design should support this integration.
- Build on the generally accepted principles of electricity system regulation: efficiency, simplicity, stability, fairness, and revenue adequacy. Therefore, net metering must balance among them.
- Be informed by the electricity system context and policy objectives related to decarbonization, equity, and system resilience, among others, and be sensitive to the locational, temporal, and scale impacts of DG. Therefore, policymakers and regulators should design net metering for the circumstances of their systems and markets, and with input from affected stakeholders.

THE REPORT

The committee recognizes that coordinating the evolution of net metering with broader electricity system changes will lead to a vastly superior outcome than addressing net metering without due consideration of its context. Therefore, the committee began its exploration by establishing a common understanding of net metering terms and its relationship to rates, how BTM DG with net metering works, and past and recent policy trends. With this background, the committee analyzed the economic, equity, technology, and regulatory considerations that policymakers should take into account as they develop and modify net metering and other policies to enable customer adoption of BTM DG and other DER. The committee then summarizes its findings and recommendations for the evolution of net metering to advance a decarbonizing, equitable, and resilient electricity system.

To begin, the committee identifies net metering as a billing mechanism used to compensate customers for the output of their BTM DG. "Traditional" net metering offsets customer electricity consumption on a volumetric (kilowatt-hour) basis with production from BTM DG, effectively crediting production at the retail rate of electricity. In recent years, policymakers have adopted several net metering "variants" that credit BTM DG production at a different rate from BTM consumption: "buy-all/sell-all" rates,[4] where consumption is charged at the retail rate paid by other customers and production is credited at what may be a different rate; use of "value of solar" approaches to calculating the credits for production; and the adoption of "feed-in-tariffs" that set credits for exports into the grid based on the costs of production or to achieve a certain level of deployment. The chapter also explains virtual net metering—where customers who cannot

[4] While the report uses the term "buy all/sell all" because it is the most common, a more accurate term may be "buy all/credit all" as customers are not generally viewed as "selling" electricity.

install BTM DG can participate in net metering by receiving credits from either off-site DG (or DG not behind their own meter), including from community solar projects, or in the context of multifamily residential buildings. Because of the integral relationship between net metering and rates, Chapter 1 offers a short primer on ratemaking to set a foundation for the discussion that follows. In addition to net metering variants (defined as rate-related modifications to net metering), the committee uses the term net metering "alternatives" that refer to DG support policies outside of ratemaking, such as tax credits.

It is also critical, from the outset, to understand how BTM DG works technically. The committee explains the basic technology by providing an example of a typical configuration for a BTM rooftop solar system. Chapter 2 provides an overview of relevant metering technologies, and concludes by describing how these systems may look when integrated with storage or other DER elements.

Having established an understanding of definitions and basic technology, the committee provides an overview of net metering in the United States, from the 1980s to the present (Chapter 3). It presents information on trends in the adoption of net metering and related policies; variants and alternatives to net metering; and deployment and costs for DG and storage. The chapter also discusses trends and tensions in the relationship between electricity market structures, net metering policy adoption, and the introduction of other DG support mechanisms, as well as related developments in energy and electricity system policy objectives.

Net metering policies raise economic, equity, technology, and regulatory, legal, and policy considerations (Chapters 4, 5, 6, and 7). The committee's recommendations recognize the importance of the availability of and access to information regarding the electricity system and BTM DG across all of these areas to enable DG to provide the greatest value to customers who adopt it, other customers, and the grid or electricity system itself. The committee also highlights the importance of taking into account the locational and temporal conditions and considerations affecting the impacts (benefits and costs) of DG on the electricity system.

More specifically, the economics of net metering and its variants are discussed from the vantage point of the customer with BTM DG participating in net metering; other non-participating electricity customers; the electricity system and the utility that serves all customers; and society as a whole, including environmental and other impacts external to the electricity system (Chapter 4).

These economics findings and recommendations recognize the importance of basing electricity prices on social marginal costs to achieve economic efficiency, while noting that other ratemaking considerations related to cost-causation, simplicity, fairness, and revenue adequacy (which includes covering the cost of the distribution system relied on even in a high DER future) may make this challenging. An alternative to changing the retail rates that underlie traditional net metering may be implementing changes to the net metering mechanism, such as net billing tariffs or buy-all/sell-all rates, with DG compensation levels set at, or near, the social marginal cost of electricity production and delivery. Where additional incentives to support BTM DG may be needed to achieve specific deployment and policy goals, the committee recommends that they be transparent and designed to achieve a specified outcome, such as encouraging BTM DG adoption by low-income customers or deployment in certain locations on the grid. The committee also notes that DG compensation should account for environmental externalities, insofar as these externalities are not otherwise internalized or accounted for.

The committee viewed examining equity considerations related to net metering as an essential part of its task (Chapter 5). Low-income households, populations of color, and renters are more likely to face barriers to DG adoption and therefore are more likely to be non-participants in net metering. As a result, economic transfers and any differential net metering benefits and costs between participating and non-participating customers have equity implications. This chapter assesses equity along multiple dimensions (distributional, procedural, intergenerational, and structural) and types (direct, indirect, and societal) in the context of access to electricity access, products and services, net metering, and DG. It discusses the interaction of net metering with inequities associated with the electricity and energy systems more broadly, including environmental, health, and resilience. The committee's findings and recommendations here emphasize the importance of attention to the equity impacts of any rate changes due to net metering for those customers least able to afford electricity. The committee further recommends that utilities make investments to increase DG hosting capacity to address any historic structural inequities in low-income communities; and the need to consider the distribution of benefits and costs (who bears what), as well as the magnitude of overall benefits and costs, associated with net metering. Beyond net metering, the committee recommends that policymakers consider and implement complementary community- and government-funded programs to expand access to BTM DG by low- and moderate-income customers. Finally, fair ratemaking should support the participation of affected customers and other stakeholders in the regulatory process so that they can have a meaningful voice.

The technology implications of increasing amounts of BTM DG for the larger electricity system underscore the need for grid modernization investments to provide visibility, integrate, manage, and/or control such DG. These investments are essential to ensure continued safe and reliable operation of the grid and provision of the grid, and to capture DG value for all customers (Chapter 6). Current and emerging advances in power electronics, storage, communications, and control technologies can enable this integration and address cybersecurity and resilience considerations resulting from BTM DG and net metering. These investments (and regulatory approval of them) need to occur simultaneously with (or even in advance of) BTM DG deployment.

The economic efficiency, equity, and technology considerations are then brought together under the umbrella of the legal and regulatory considerations that affect the pace and character of the adoption and evolution of net metering policies across states and utilities (Chapter 7). The committee describes how regulatory and legal considerations interact with net metering policy related to technology innovation and deployment that can provide support for a decarbonizing, equitable, economically efficient, and resilient electricity system. The discussion includes the jurisdictional split between wholesale (federal) and retail (state) regulation of electricity and the status of federal regulation of net metering, the intersection of net metering policy with traditional electric utility ratemaking, the utility's obligation to serve, and other policies pertaining to the evolving relationships between utilities and their customers including those who adopt technologies that provide part of their power needs.

The committee emphasizes the importance of examining net metering policies, taking into account the current technology costs and market maturity of DG, both for states that have previously adopted net metering and are considering variants, as well as for parts of the country that have not yet adopted net metering. Policymakers and regulators should recognize traditional ratemaking principles as updated to reflect the application of new technologies and service offerings, including cost-causation, rate simplicity, fairness, revenue adequacy, and compensating resources based on their value. Decision makers should design compensation approaches for the export of power from BTM DG consistently with how the utility values other sources of power that offer comparable services to the system. Finally, given the economic and equity challenges associated with using net metering and other incentives and programs where costs are recovered in electricity rates, policymakers should consider net metering alternatives such as tax incentives, building codes, and financing assistance to low-income households.

The report concludes by highlighting the success of net metering to enable and support the deployment of BTM DG, advancing policies to reduce climate impacts, meeting customer expectations for better products and services, and encouraging technological innovation (Chapter 8). Thanks to rapid technological advancements, cost reductions, net metering, and other policies, DER generally, and BTM DG (especially distributed solar) more specifically, are playing a more central and active role in the electricity system than was thought possible a couple of decades ago. Moving forward, to enable the integration of BTM DG into the electricity system to maximize benefits for all customers, traditional net metering needs to be revised to align value streams, costs, and compensation.

It is critical to ensure that revisions to net metering also improve equity, an outcome that must be achieved through a continual, intentional approach. Fully integrating BTM DG into the electricity system portends a more resilient electricity system, and achieving this goal entails systematic, coordinated, and sustained investments across distribution systems. Policies and regulations must support the creation of a framework and market conditions that enable utility and, as applicable, non-utility participants to make these investments in the distribution. Finally, broader societal goals may be best pursued through approaches and mechanisms that go beyond ratemaking to achieve key objectives, especially equity. With a more open, transparent, deliberate, and intentional approach that leverages the locational, temporal, and contextual value streams of BTM DG, an evolved net metering policy could enable an electricity system that is more sustainable, equitable, and resilient. The report ends with a full listing of all of the recommendations from the chapters.

1

Introduction

Net metering (net energy metering or NEM) is a policy adopted in many states to encourage the deployment of clean, distributed generation by residential, commercial, and industrial electricity customers, on the customer side of the meter known as "behind the meter" (BTM). Policymakers implemented NEM to help achieve clean energy resource diversity, financing assistance, carbon reduction goals, and in some instances to encourage economic development, increase resilience, and reduce system costs.[1] For example, Rhode Island's net metering law states that its purpose is:

> to facilitate and promote installation of customer-sited, grid-connected generation of renewable energy; to support and encourage customer development of renewable generation systems; to reduce environmental impacts; to reduce carbon emissions that contribute to climate change by encouraging the local siting of renewable energy projects; to diversify the state's energy generation sources; to stimulate economic development; to improve distribution system resilience and reliability; and to reduce distribution system costs.[2]

Net metering is receiving increasing amounts of attention, support, and scrutiny today as greater quantities of distributed generation (DG) and distributed energy resources (DER) are connected to the electricity grid, spurred not only by policy but also by economic, technology, and social developments. Net metering policies have been a key determinant in the adoption of DG and have generally succeeded in encouraging deployment of BTM clean DG, primarily rooftop solar. The technical requirements for an electric power system with these distributed elements are detailed in Chapters 2 and 6. Broad decarbonization and resilience objectives, as well as economic development and job growth goals, are also driving increasing interest and uptake in distributed and grid-scale technologies for renewable electricity generation, grid-connected storage, and demand management. This growth in DG and the number of assets connected to the electricity grid that are subject to net metering and related policies highlight the need to understand the economic, equity, environmental, and technology implications of net metering to help decision makers make informed choices about this and related policies in the future.

---

[1] Wan, Y.H., and H.J. Green. 1998. "Current Experience with Net Metering Programs." CP-500-24527. Golden, CO: National Renewable Energy Laboratory. Paper presented at WINDPOWER '98, Bakersfield, CA. https://www.nrel.gov/docs/legosti/old/24527.pdf.

[2] Rhode Island General Laws. "Title 39, Chapter 26.4, Net Metering." State of Rhode Island General Assembly. http://webserver.rilin.state.ri.us/Statutes/TITLE39/39-26.4/39-26.4-1.htm.

The committee's statement of task (Box 1-1) asks it to evaluate NEM's potential to contribute to a decarbonizing, equitable, and resilient electricity system in the context of alternative transactional mechanisms and incentives. The statement of task further instructs the committee to assess the "medium-to-long term impacts of net metering on the electricity grid and consumers." In response to this charge, the committee has developed this report which presents guiding principles for rate design, technology, and regulation, among other relevant factors in order to inform prudent, adaptable, and forward-thinking policy for net metering, its variants, and alternatives. To begin, the committee considers "medium-term" to encompass a time period of approximately 5 to 10 years, where there is some foresight of changes in conditions, technologies, and policies affecting net metering and DG deployment. The "long-term" refers to 10 to 20 years or more following this report's publication (in 2023), where additional changes can be expected but their nature and scope cannot yet be fully foreseen.

The committee also considered near-term changes that are already under way or in process. The evolution of net metering policy can be viewed as corresponding to the growth in customer adoption of BTM DG. In the early stages, the objective of net metering was to encourage adoption. The incentives it provided had little impact on customer rates, including "non-participants" (i.e., those customers who did not adopt BTM DG). As deployment of BTM DG increased, net metering policies were modified – both to support DG growth by lifting caps on eligibility and by modifying the incentives

BOX 1-1
STATEMENT OF TASK

In its 2020 appropriations for the Department of Energy, the U.S. Congress directed the National Academies of Sciences, Engineering, and Medicine to appoint an ad hoc committee of experts to study the issues associated with net metering, including the medium-to-long term impacts of net metering on the electricity grid and consumers. This study will begin with a definition of net metering and a discussion of the history and implementation of net metering. The study will include a discussion of net metering planning and operating techniques, business models, and architectures, as well as electric service rate designs and new technology developments. Current net metering and alternative policies will be discussed as distinct topics because each may have different underlying enabling technologies. The committee will provide an analysis of the opportunities and challenges relevant to net metering, including broad stakeholder acceptance and adoption of distributed renewable energy resources. The committee will assess net metering in the context of alternative transactional mechanisms and incentives, and evaluate its potential to contribute to a decarbonizing, equitable, and resilient electricity system. The study will give recommendations on key principles for policymakers when considering net metering and alternative policies.

In developing its report, this committee will consider the following key topics.

- Performance and Effects of Net Metering and its Alternatives on Business Models, Consumers, Technology Suppliers, Electricity Markets, and Grid Operations at High Penetrations of Distributed Renewables. The committee will examine the capability of net metering and successor policies to provide accurate valuations of electricity, ensure sufficient financial support for investment and operations necessary for the system, and support equitable, well-functioning, and economically efficient markets as the quantities of renewable resources and other assets connected to the distribution side of the grid greatly increases. Metering alternatives should be described and assessed on their potential to provide improvements in transactions, measurements, and incentives for the use of distributed renewable generation. As new metering technologies allow for more complexity in valuation and billing, the ability of consumers to understand this complexity will be considered by the committee.
- Emergence of New Technologies, and Effects on Consumer and Industry Transactions and Incentives. The committee will consider how new technologies and their attendant new capabilities alter the roles of the participants in the system, the requirements of each participant to fulfill those roles, and the challenges and opportunities present. Consumers in particular will need the information and context necessary to participate fully. Customer-owned distributed energy resources are capable of providing grid services to both distribution and transmission entities in addition to any purpose a customer may choose, and therefore issues of consent, compensation, and privacy should be studied by the committee.
- Alternative Metering Systems that Promote Equitable Distribution of System Resources and Costs. Net metering and supporting technologies should be considered both from the perspectives of participants and non-participants, and key principles for assessing fairness should be discussed. Fair and equitable electric service, which is an ultimate system objective, is not addressable solely by net metering or any successor policies. As such, the discussion should be within the broader context of electric utility regulatory policies that impact equity, including access to distributed energy resources and designing successor tariffs to increase the adoption of distributed energy resources by customers from lower-income and marginalized communities.
- How Net Metering and Alternatives May Affect the Evolution of the Electricity System. This study will assess the extent to which net metering and alternative policies may adapt to or enable different electricity system structures, such as one with increased distributed elements for advanced reliability and resilience, equity, and the adoption of different technologies, such as behind-the-meter storage systems and rooftop and community solar.

provided to reduce their impact on rates. As DG adoption continued to grow, net metering policy variants were introduced, including time-varying and other modifications to rates to better align incentives with the value BTM DG could provide. As this occurred, the objective of net metering policy began to evolve toward facilitating optimization of system design encouraging DG more in certain locations and at certain times than others. Looking forward as the electricity system itself continues to evolve with increased focus on decarbonization, equity and resilience, net metering policy should continue to evolve with it toward a sustainable structure that benefits the system and all customers (see Figure 1-1). The shift to pricing based on avoided costs shown in Figure 1-1 has ranged from wholesale prices to estimates of the value of DER to the system, including externalities; Chapter 4 describes the elements involved in estimating that value.

NET METERING DEFINED

The committee recognizes that the term "net metering" is often employed with varying intended meanings and assumptions. To limit confusion and potential misunderstanding

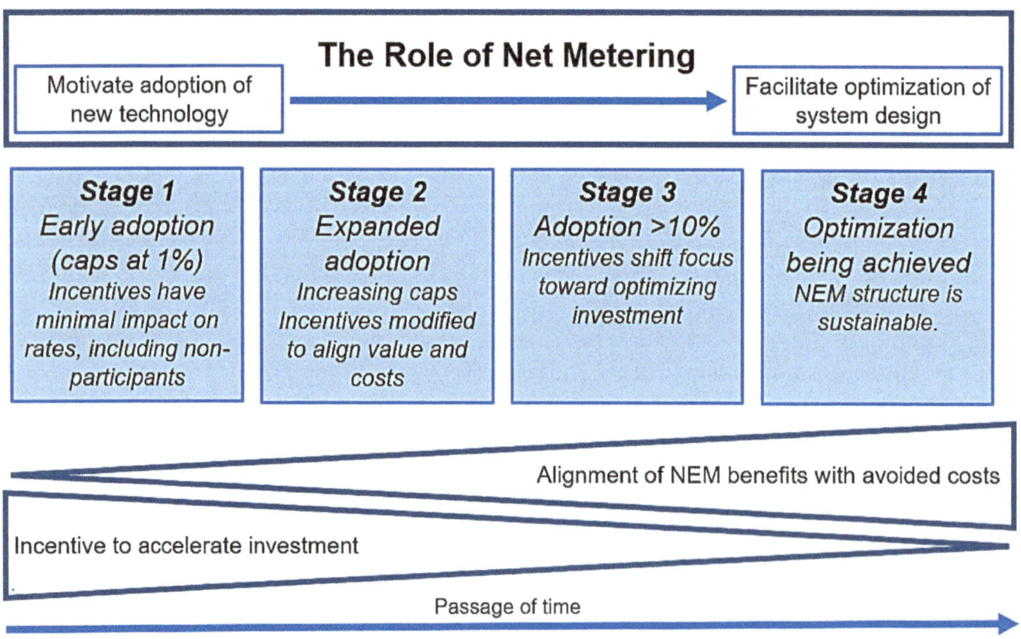

FIGURE 1-1 A visual depiction of the evolution of net metering within the context of increasing behind-the-meter distributed generation and corresponding evolution in the electricity system over time.
SOURCE: Adapted from S. Kurtz.

in the context of this report, the committee presents the following definitions of the term "net metering" and its variants, reflecting the most-common meaning of these terms.

The committee defines net metering as a *billing mechanism* that (1) offsets customer electricity consumption on a volumetric (kilowatt-hour [kWh]) basis with production from BTM DG, effectively crediting production at the retail rate applicable at the time of consumption; and (2) similarly credits customers at the retail rate for electricity exported to the electricity grid. As initially implemented using a single, analog utility meter for each participating customer, a utility customer's billing meter ran backward as electricity was exported to the electricity grid, and forward as electricity was consumed from the grid.[3]

This definition of *net metering* refers to traditional net metering, where at the end of each billing period the net amount of energy (measured in kWh) consumed or exported to the grid would be recorded. The customer would either owe a payment to the utility for the net energy consumed, or a credit would be calculated, based on the net energy exported to the grid. Such a credit might carry forward for application in a subsequent month. In traditional net metering program designs, participating customers might be eligible for payment for any net excess generation, usually at the end of a year or the end of 12 consecutive billing cycles, with payment for this generation credited at either the full retail rate for consumption or another rate. Implementation details have varied by jurisdiction. In some cases, program design differences reflected variations in the capabilities of pre-existing utility metering and billing systems.

The committee notes that the time period over which credits are calculated (or netted) can affect the level of compensation for customers who are producing or generating electricity and participating in net metering. Other policy choices or conditions on net metering can also affect the compensation provided to customers for their BTM DG. These can include:

- Limits on the size of systems eligible for net metering (e.g., no more than 125 percent of customer load) to encourage matching the DG capacity to historic or anticipated levels of customer consumption.
- Caps on the overall amount of generation eligible for net metering on a utility's system (e.g., number of megawatts [MW] or a percentage of a utility's overall

[3] Adapted from Gilliam and Stanton, p. 9. Stanton, T., and D. Phelan. 2013. "State and Utility Solar Energy Programs: Recommended Approaches for Growing Markets." NRRI 13-07. Silver Spring, MD: National Regulatory Research Institute. https://pubs.naruc.org/pub/FA86BBED-D69F-C1EC-D308-0BB2CDAF5E37. Quoting from Gilliam, R. 2013. "Net Metering." p. 5, PowerPoint presentation. *The Cost, Benefits and Equity Issues of Net Energy Metering*. February 5. Washington, DC: NARUC Winter Committee Meetings.

load) to reduce the potential impact of increasing amounts of DG on the electricity system.
- The ability to cash out credits at the end of the netting period will also affect the level of net metering compensation.

The level of compensation also affects non-participating customers and the host utility. Timing of output from on-site generation can affect the electricity system (distribution, transmission, and generation), particularly where the timing of production does not match the timing of use by others on the electric system. The location of distributed generation (e.g., on constrained distribution circuits) can also affect the electricity system. Some of these impacts may lead to costs to the system and other customers (e.g., where DG increases constraints), and other impacts may result in assistance to the grid and benefits to other customers (e.g., where DG relieves constraints). These impacts, and methods for influencing them, are examined in this study. (See Chapter 4 for a discussion of economic impacts and Chapter 6 for a discussion of physical and technological impacts.)

In addition to traditional NEM, state policymakers and regulators have developed and are exploring NEM variants and alternatives that may better accomplish decarbonization, equity, and resilience objectives. This report uses the term *net metering variants* to refer to refinements to the ratemaking approach or design that underlies traditional net metering and the term *net metering alternatives* to refer to policies and programs outside ratemaking that have been implemented or are under consideration. The report uses the term *net metering and its variants and alternatives* when referring to traditional net metering, ratemaking-related refinements, and alternatives to it.

NET METERING VARIANTS

Variants for Customers with On-Site Generation

Prominent examples of net metering variants that largely rely on the underlying utility ratemaking or rate design structure include the following:

- **Net Billing:** A tariff designed to compensate customers with distributed generation for exported production at a netting interval (a defined time period) that may be shorter than the usual monthly billing period (e.g., daily, hourly, time-of-use window, or even instantaneously) and/or at a rate which may differ from the retail rate (e.g., an administratively determined avoided cost rate or wholesale market price; essentially netting charges and credits on the bill rather than imports and exports during a specific time interval). With this type of net metering, customers may not be able to use credits accrued during one time period in a future billing cycle or apply them to all elements on the bill, such as customer or system benefit charges.

- **Inflow/Outflow Rates:** A variation on net billing where a customer avoids paying for electricity at the retail rate when the customer's own generation is consumed onsite and offsets that consumption, but outflow to the grid is credited at a rate different from the retail rate.
- **Buy-All, Sell-All Rates:** Customers with distributed generation purchase electricity from the utility at the same rate as those without, and sell electricity back to the grid at a rate that could be administratively determined, set at avoided cost or wholesale market price. With buy all, sell all rates, customers are not treated as if they directly consume any of the energy generated by their solar or other BTM DG.

Two common ways to set the price or compensation level for solar production under Net Billing and Buy-All, Sell-All Rates include administratively determined rates intended to reflect the value to the electric utility system or feed-in tariffs (FIT) intended to produce a reasonable return on investment based on the installed cost of DG and its production. Administratively determined rates range widely in different service territories, from average annual wholesale prices for energy at the low end, through prices reflecting the estimated value of solar (VOS) or value of distributed energy resources (VDER) to the system (including externalities) at the high end.

- **Value-of-Solar (VOS, or Value of DER [VDER]) Tariff:** Customers are compensated for their behind-the-meter generation at a rate based on the value of the solar (or other DER) to the utility, the distribution grid, the electric system, and/or society, rather than the retail rate. The VOS rate may be applied to (1) all generation, including both generation that offsets consumption and generation that is exported; or (2) only exported generation. The value of solar, or method to calculate it, is generally determined through an administrative proceeding.
- **Feed-In Tariff (FIT):** Customers are compensated for generation sold to the grid under a tariff at prices set administratively, which may be intended to achieve certain deployment goals or based on cost estimates.

This list is not exhaustive but instead represents prominent net metering variants. A visual depiction of the relationship between net metering, its variants, and their definitions is included in Figure 1-2.

TRADITIONAL "NET METERING" *a* ***billing mechanism*** *that (1) offsets customer electricity consumption on a volumetric (kilowatt-hour [kWh]) basis with production from behind-the-meter distributed generation, effectively crediting production at the retail rate applicable at the time of consumption; and (2) similarly credits customers at the retail rate for electricity exported to the electricity grid. As initially implemented using a single, analog utility meter for each participating customer, a utility customer's billing meter ran backward as electricity was exported to the electricity grid, and forward as electricity was consumed from the grid.*

PROMINENT NET METERING VARIANTS

NET BILLING

A tariff designed to compensate customers with distributed generation for exported production at a netting interval (a defined time period) that may be shorter than the usual monthly billing period (e.g., daily, hourly, time-of-use window, or even instantaneously) and/or at a rate which may differ from the retail rate (e.g., an administratively determined avoided cost rate or wholesale market price; essentially netting charges and credits on the bill rather than imports and exports during a specific time interval). With this type of net metering, customers may not be able to use credits accrued during one time period in a future billing cycle or apply them to all elements on the bill, such as customer or system benefit charges.

BUY-ALL/SELL-ALL RATES

Customers with distributed generation purchase electricity from the utility at the same rate as those without, and sell electricity back to the grid at a rate that could be administratively determined, set at avoided cost or wholesale market price. With buy all, sell all rates, customers are not treated as if they directly consume any of the energy generated by their solar or other BTM DG.

INFLOW/OUTFLOW RATES

A variation on net billing where a customer avoids paying for electricity at the retail rate when the customer's own generation is consumed onsite and offsets that consumption, but outflow to the grid is credited at a rate different from the retail rate.

VALUE-OF-SOLAR (VOS, OR VALUE OF DER [VDER]) TARIFF

Customers are compensated for their behind-the-meter generation at a rate based on the value of the solar (or other DER) to the utility, the distribution grid, the electric system and/or society, rather than the retail rate. The VOS rate may be applied to (1) all generation, including both generation that offsets consumption and generation that is exported; or (2) only exported generation. The value of solar, or method to calculate it, is generally determined through an administrative proceeding.

FEED-IN TARIFF

Customers are compensated for generation sold to the grid under a tariff at prices set administratively, which may be intended to achieve certain deployment goals or based on cost estimates.

FIGURE 1-2 A visual depiction of net metering, its prominent variants, and their relationships to each other.
NOTE: Lines indicate direct relationships between particular net metering variants, which may build off each other.

Virtual Net Metering

Another approach to NEM is **virtual net metering (VNM).** Virtual net metering extends the net metering billing and crediting mechanism beyond the meter of the location where electricity is produced to allow the assignment of credits to customers in another location. VNM distributes the credits for DG production among multiple owners or subscribers, or with multiple service addresses for a single utility customer, regardless of whether the DG is co-located with every, or any, member. For example, a solar developer could build a project (sometimes characterized or structured as a *community solar* project) and assign net metering credits to several customers, for instance in a housing project.[4] Another form of virtual net metering is when a business, university or hospital campus, or municipality installs a DG project behind the meter at one building or location and assigns credits to other buildings or municipal facilities with their own meters. Under virtual net metering, credit value may be set as it is for net metering generally: (1) at the retail rate; (2) at an administratively determined rate (which can range from an avoided cost rate, a wholesale market price, or a VOS rate); or (3) at a rate, such as a FIT, intended to produce a reasonable return on investment for the installed equipment.

Virtual net metering evolved to enable customers to participate in DG adoption when they do not have suitable conditions for installing DG behind their own meters. For example, their homes or buildings may be heavily shaded and not appropriate for rooftop solar, or they may live in apartment buildings with one meter. VNM enables them to essentially "use solar PV even though it is installed on another building or site." Through VNM, these customers receive credits on their bills that reflect the energy the DG system generates and provides to the grid. When credited at the retail price for billing purposes, it is as if the customer's electricity meter is running backward.

There are two primary types of virtual net metering. The first is where a single organization or customer owns both assets (the DG and the load). For example, a municipality may have a solar PV farm on its landfill and use the credits generated to offset the bills of municipal buildings and streetlights on their accounts with their electric utility. In this type of virtual net metering all the meters for the systems owned by a single entity are summed (with DG generation being negative) and the municipal customer pays the net amount used.[5]

The second type of virtual net metering allows for multiple customers to offset their electricity bills using a separately located and shared DG system. An example of this approach would be a community solar project. A community solar project could be owned by the

[4] Some jurisdictions call this "remote" net metering.

[5] Some jurisdictions call this "aggregate" net metering.

electric utility, which would provide customers with local green power or credits from the solar facility. The credits from the virtually net metered solar system reduce customers' bills without their having to incur the generally higher cost of installing and operating their own rooftop solar system. In this configuration, the utility generally owns the solar project and continues to generate the electricity provided to customers. Community solar projects can also be built and owned by solar and storage developers who assign net metering credits to customer "off-takers." In this situation, the electricity is generated by projects not owned by the utility, reducing its sales to customers.[6]

NET METERING ALTERNATIVES

Net metering (including its rate-related variants) is one way to encourage the deployment of BTM DG. There are also several alternative policies and mechanisms, designed to encourage the deployment of distributed clean generation, which go beyond compensation through utility rates. They include:

- Renewable portfolio standards (RPS) or clean energy standards (CES), particularly when specifying quantities or percentages for distributed generation
- Rebates or financial incentives funded outside of utility rates, for example through federal or state budgets that authorize and appropriate funds to programs (Note: does not include utility program funds which would be recovered from ratepayers)
- Tax credits (state or federal, or both) for use by individual households and businesses
- Building codes (e.g., that require the installation of rooftop solar on certain new construction)

This report will discuss some of these alternatives in relation to net metering, where they may complement, substitute for, or enhance the efficiency and effectiveness of net metering and its variants.

RATE DESIGN AND NET METERING

As noted in the definition above, net metering is a billing mechanism intrinsically tied to the retail electricity rate structure. Compensation for BTM DG provided by net metering

[6] Community solar owned by entities other than the utility can have similar impacts to individual rooftop solar on the utility system and for other customers. These impacts are discussed in Chapter 4 (Economic Considerations Related to Net Metering).

can and has been set and modified through changes to the underlying rate structure independently of any changes to net metering per se. (See Chapter 3 discussion.) Refinements or changes to retail rate design will affect the level of the credits (or compensation) net metering provides to BTM DG customers, as well as impacts on other customers and the electricity system. These refinements involve (1) changes to the structure of rates, including the level of monthly customer charges or other fixed charges; (2) changes in energy charges, such as time-of-use or other time-varying rates for consumption (cents per kWh); and (3) a charge tied to the level of the customer's peak demand during a period; and/or mechanisms such as minimum bills. To understand the implications of these changes on customers with and without BTM DG, as well as the utility, it is important to understand how rates are set.

Mechanics of Traditional Ratemaking and Rate Design

There are multiple goals of traditional utility regulation and ratemaking, including revenue sufficiency and stability for utilities, consumer protection, understandability of rates and bills, administrative feasibility, and pricing efficiency. Regulators must balance among them, using their judgment. (See Chapter 7 for further discussion.) That said, traditional utility ratemaking begins with a calculation of what it costs the utility to provide service to customers, including both embedded capital investments, and expenses such as operations, maintenance, and fuel. The total cost of service is also referred to as a utility's revenue requirement. The revenue requirement is then translated into the rates customers pay so that the utility is given a reasonable opportunity to recover the costs it prudently incurs to provide service to customers. This is a multi-step process:

- First, costs in a particular time period (e.g., a year in the past or a forecasted year in the future) are identified by function, such as generation, transmission, distribution, and customer connection.[7] Some of the costs may be fixed over a period of time (such as the depreciation and interest expenses for the capital

[7] Note that this description of ratemaking mechanisms reflects the costs of a vertically integrated utility that provides bundled generation, transmission, and distribution service. Where a utility provides distribution service alone and either purchases generation and transmission service from an all-requirements supplier, that utility's own rates reflect their costs of the distribution function (and generation and transmission costs they pay are passed through to customers as part of the electricity bill). Similarly, in parts of the country where a customer can elect to have a competitive supplier other than the utility provide commodity supply, the distribution utility's rate will reflect its costs to provide "wires" service (its own distribution costs and its own or passed-through transmission costs), and the customer's bill will typically reflect the passed-through price of the competitive supplier. See Chapter 7 (Regulatory, Legal, and Market Considerations) for a discussion of organizational types of utilities.

costs of a generator, or distribution and transmission investment, or labor costs over a year). Other costs are variable, changing with usage (such as fuel and some operating and maintenance expenses).

- Next the cost of service is allocated to rate classes (based on the different types of costs to serve them). Rate classes are similarly situated customers; similarly situated means that on average across a group of customers their usage levels and patterns are roughly similar. Common rate classes are residential, commercial, and industrial. Any individual's usage may differ from the average of its class.
- Rates are then designed to collect the costs to serve from these different groups of customers based on projections or estimates of their class consumption and other factors.
- Historically, rate design recovers different types of costs through different types of charges or billing components on the customer bill:
  - *Customer charges* ($/month) cover the costs to connect a customer to the electricity system. Traditionally, these include the service drop (wire from distribution system to house, commercial building, or industrial facility) and costs associated with billing, metering, and collections. In some jurisdictions, the customer charge includes a minimum level of distribution investment needed to provide service to a customer. Recently, other fixed monthly charges have been considered or included in customer bills.[8]
  - *Energy charges* (cents/kWh) cover the costs that vary with consumption or usage, such as fuel costs associated with generating the kWh consumed. In rate designs without demand charges or customer charges reflecting only the costs to connect, energy charges have also typically included what may be considered fixed (capital) costs of generation, transmission, and distribution, which have been amortized over a period representative of the life of the investment. (This can be thought of as similar to a bakery including the relatively fixed costs of its building, ovens, and other equipment, in addition to the costs of flour, eggs, and butter, which clearly vary with the volume of baked items provided and sold.)[9]

---

[8] Prehoda, E., J.M. Pearce, and C. Schelly. 2019. "Policies to Overcome Barriers for Renewable Energy Distributed Generation: A Case Study of Utility Structure and Regulatory Regimes in Michigan." *Energies* 12(4):674.

[9] In restructured states, the utility bill may include supply charges separately from transmission and distribution charges. These supply charges include the generation or energy purchased to serve customers. The transmission and distribution charges include the costs of poles and wires and other transmission and distribution equipment. All are usually billed on a volumetric basis (characterized as an energy charge) rather than as a "demand" charge.

  - *Demand charges* ($/kW) cover the costs associated with putting in place the capacity needed to provide the maximum amount of power a customer may consume at a defined point of time (e.g., a 15-minute interval) in a month. Historically, residential and small commercial customers' rates have not included demand charges; these demand charges have been part of large commercial and industrial rates. Demand charges recover a portion of fixed charges associated with capacity costs (i.e., investments in generation, transmission, and distribution) to ensure that the utility has the capability to meet customers' peak demand even if that capacity is not used every hour. While demand charges vary with peak usage, they may be considered fixed charges for some period (e.g., during a period established in the tariff that sets the demand charge equivalent to a customer's highest usage during a month, season, or year, after which the demand-charge level may be reset to reflect the customer's new peak usage during that operative period).
  - Other charges on a customer's bill might include a *system benefit charge* to recover the costs of utility energy efficiency programs, low-income discounts (from non-low-income customers), or other mandated public policies. System benefit charges are generally assessed on a cents/kWh basis but are also sometimes included in the $/month customer charge. System benefit charges are sometimes included in the rates for all customers, even where customers have the option to purchase generation service from an entity other than the utility.
- Once costs are assigned to a category, they are divided by a forecast or estimate of the number of units expected to be sold in that category (billing determinants) in order to determine the rate. The billing determinants are the number of customers for the customer charge, kWh of usage for the energy charge, and kW of demand (peak usage at a point in time) for the demand charge.

Decisions about whether to recover costs in customer, energy, or demand charges are policy choices based on how customers use the electricity system, the costs incurred to serve those uses, and other factors—such as, customers' ability to respond to and modify their usage in response to the price signals sent by rates. (For example, large electricity users, particularly large commercial and industrial customers, have historically been more sophisticated with respect to billing and energy management, especially where electricity costs are a significant portion of their budgets.) While kWh usage/consumption and costs are clearly interrelated in some instances (e.g., consumption levels driving fuel costs associated with producing power), the association between consumption and costs is less clear in other instances (e.g., usage levels that affect distribution investments which tend to be relatively capital intensive and thus fixed over a period).

Mechanics of Net Metering and Variant Options

Under traditional net metering, customer-sited generation offsets usage or consumption, effectively crediting the generation at the retail energy rate (cents/kWh) and excess generation exported to the grid is similarly credited at the retail rate. In a number of jurisdictions, customers can use these credits to offset their customer charge ($/month) and any other charges on their bill, further reducing payments they would have otherwise made to the utility. In some jurisdictions, under traditional net metering, the credit for BTM DG production may be reduced by system benefit charges so that it is not equal to the full retail rate, with the intent that net metering customers contribute to paying for these system benefits.

As customer electricity usage and, in the case of BTM DG, production changes the utility's cost recovery will change absent a change in rate structure, both in terms of the amount collected and from which customers it is collected. If a customer reduces their consumption from the utility because they install solar (or undertake energy efficiency), the utility's revenues (and cost recovery)[10] will similarly be reduced. With the increasing deployment of BTM DG under traditional net metering, customer consumption, and therefore utility revenues, have been declining.[11] There are a number of rate design options to address declining (or increasing) cost recovery. Ideally, increased BTM DG has also decreased the utilities' costs. In most cases, the change in the costs have not matched the changes in the revenues from the associated customers, creating a new challenge in the standard calculation of rates needed for cost recovery.

- *Decoupling:* Decoupling is a mechanism that utilities can use to adjust rates to align cost recovery between rate cases and has been widely adopted related to energy efficiency.[12] It de-links or "decouples" revenues from sales,[13] allowing

[10] While increased BTM DG may also decrease the utilities' costs, in many cases, the change in costs has not matched the change in revenues from the associated customers.

[11] Conversely, other changes in the evolving electricity system, such as electrification of transportation and buildings, will lead to increased consumption and utility revenues. This suggests that any rate mechanism to account for consumption levels that differ from forecasts should operate in both directions.

[12] Subramanian, S., W. Berg, E. Cooper, et al. 2022. "2022 State Energy Efficiency Scorecard." *American Council for an Energy-Efficiency Economy*. https://www.aceee.org/sites/default/files/pdfs/u2206.pdf. See pp. 46, 49–51, table by state including both electric and natural gas decoupling yes or no for each state.

[13] Ratemaking approaches such as performance-based regulation can also be viewed as de-linking revenues from sales because they base rates or adjustments to rates on outputs or performance, unlike traditional regulation that bases rates on inputs or costs.

rates to adjust with changes in sales to achieve a revenue level previously approved in a rate case. When usage or sales are declining, these adjustments increase the cost per kWh and can have a differential impact on bills for different customers. Customers without BTM DG will experience increased bills (all else equal) compared to customers with BTM DG. (In fact, under traditional net metering, customers with BTM DG will see an increase in their credit value as rates increase.)

- *Net Billing*, with reduced credits for the exported generation: Another way to reduce the impact on utilities of reduced revenues resulting from net metering BTM DG is to credit excess generation exported to the grid (or *outflow*) at a rate that differs from the retail rate. This rate could be administratively determined, based on avoided costs, wholesale market prices, or the *value of solar*. Depending on how avoided costs are calculated (including wholesale, retail, and other utility costs) to arrive at a credit, then rates and costs could be aligned. An administratively determined rate or wholesale market price that is less than retail rates will reduce the cost to the utility of net metering credits. If the value of solar is less than the retail rate, the cost of credits to the utility will be reduced. If the value of solar includes societal costs (e.g., health and climate benefits) that are external to the utility, its cost of credits may increase. (See Chapter 3 for a discussion of net billing adoption and Chapter 4 for a discussion of impacts on the utility, society, and customers with and without DG.)
- *Increased Customer* or *Other Fixed Charges:* As an alternative to increasing energy rates, utilities can mitigate the cost recovery effect of declining usage by recovering more of their costs in another charge on the bill. Options include the monthly customer charge or another new fixed charge. In states with significant increases in BTM DG, these options have received increasing attention in net metering policy discussions. (They are discussed further in Chapter 3.)
- *Minimum Bill:* An alternative to an increased or new fixed charge is a minimum bill. Customers would pay a minimum amount to the utility every month regardless of their level of consumption. The level could generally be set such that average monthly bills for customers without BTM DG would exceed the minimum so that they would not see bill increases. For customers with BTM DG, the minimum bill might exceed their new level of usage. To the extent that customers cannot use credits from net metered BTM DG, the credits could be carried over to future months or cashed out at the end of a year or another time period. Minimum bills can be tiered by average usage level or income to ensure that they do not increase costs for low usage, low income, or otherwise disadvantaged customers. (See Chapter 5 for further discussion of the importance of addressing these equity concerns.)

- *Time Varying* or *Time-of-Use Rates:* To the extent that utility costs vary by time of day or season of the year, the energy charge can be based on time of use to better align rates with costs. With time-varying rates, customers who reduce their consumption through BTM DG and export excess generation to the grid during time periods with high costs may not reduce utility cost recovery to the degree that they would with flat energy charges because the reduction in utility revenues is aligned with the reduction in utility costs. Consequently, the utility would not have to increase rates to maintain revenues, which can shift costs from customers with BTM DG to customers without. (See further discussion in Chapters 4 and 5.)
- *Standby* or *Grid Access Charges:* Another rate design option that introduces fixed charges applicable only to customers with BTM DG is a standby or "grid access" charge. It is a fixed monthly charge for the ability to connect to and rely on electricity from the grid when BTM DG is not producing and results in revenues to the utility to offset those lost due to reduced usage or consumption. Standby charges have been applied to large commercial and industrial customers with their own generation (often cogeneration) for some years. Grid access charges have been proposed more recently in nine states but have generally not been implemented. (See Chapter 3 for further discussion.)

Further Considerations Affecting Rate Design

Setting rates to recover utility costs of service is a key principle guiding ratemaking. It can help to avoid negative impacts and inaccurate price signals for all customers. However, as noted earlier, rate design requires balancing several considerations in addition to cost recovery, including rate simplicity, fairness, and gradualism, among others. These are discussed further in Chapter 7 and should be considered as policymakers look at net metering and ratemaking variants. Net metering alternatives, outside of ratemaking, also warrant consideration.

REPORT STRUCTURE

Based on the committee's statement of task, this report addresses numerous issues associated with net metering. These include trends in the deployment and costs of on-site DG and the use of net metering policies; economic, equity, technology, and regulatory policy considerations relevant to net metering and its variants; and the committee's perspectives on the roles that net metering, its variants, and other alternative policies may play in the future.

The committee recognizes that coordinating the evolution of net metering with broader electricity system changes will lead to a vastly superior outcome than addressing net metering without due consideration of its context. Therefore, the committee began its exploration by establishing a common understanding of net metering terms and its relationship to rates, how BTM DG with net metering works, and past and recent policy trends. With this background, the committee analyzed the economic, equity, technology, and regulatory considerations that policymakers should take into account as they develop and modify net metering and other policies to enable customer adoption of BTM DG and other distributed energy resources (DER). The committee then summarizes its finding and recommendations for the evolution of net metering to advance a decarbonizing, equitable, and resilient electricity system. Briefly,

- Chapter 1, Introduction, begins with the committee's statement of task, definitions, and basic ratemaking.
- Chapter 2, Net Metering 101, explains the mechanics of the system to the lay reader.
- Chapter 3, Background and History, Current Status, and Near-Term Future of Net Metering, lays out the current state of and trends pertaining to DG technology adoption, costs, and net metering policies.
- Chapter 4, Economic Considerations Related to Net Metering, addresses economic and rate design considerations.
- Chapter 5, Equity Considerations of Net Metering, discusses the equity outcomes associated with net metering.
- Chapter 6, Net Metering and Distributed Energy Technologies, then provides an overview of the effect on the grid of relevant technologies influenced by net metering and grid technologies to integrate, manage, or control them.
- Chapter 7, Regulatory, Legal, and Market Considerations, presents a discussion of the legal and regulatory context and considerations for net metering policy.

The committee presents possible futures and outcomes and lists all recommendations from the report in Chapter 8, The Future of Net Metering in an Evolving Electricity System.

2

Net Metering 101

Chapter 1 describes net metering as a compensation mechanism and lays out key definitions for understanding net metering. This chapter describes the basic technology behind how net metering for behind-the-meter (BTM) distributed generation (DG) works. This discussion begins by presenting a typical configuration for a BTM rooftop solar system, then provides an overview of relevant metering technologies, and concludes by describing how these systems may look when integrated with storage or other distributed energy resources (DER) elements, or both.

TYPICAL CONFIGURATION FOR A ROOFTOP SOLAR SYSTEM

All electricity users or consumers (e.g., residential customers in a single-family home or apartment building, commercial customers in schools or businesses, and industrial customers with factories) that are connected to the grid have a meter to record (i.e., measure) the amount of electricity they use. There are many types of metering technologies and equipment. Historically, residential and small commercial customers' electricity usage was recorded by a simple meter that measured the amount of electricity drawn or consumed from the grid over the course of a billing period, usually a month (measured in kilowatt/hour [kWh]), without regard to when and how much electricity was used at different times during that month. Larger users (e.g., a large commercial office building or an industrial facility) had meters that recorded not only monthly consumption but also the maximum instantaneous draw from the grid during a month, or the peak use each day (measured in watts or kilowatts), or some other dimension that examined demand levels at different points of time during a period (e.g., watts of demand on a Thursday between 5:00 and 5:05 pm on a hot day). In recent years, metering technology has become much more sophisticated, enabling data recording and communications of the customer's usage across the hours of a day and across the days of the billing period.

With the proliferation of BTM DG, metering technologies have also become more sophisticated to accommodate and support policies like net metering, which encourages DG adoption. When it comes to the interconnection of DG with the electricity grid, the most common type of household DG for much of the world is a solar photovoltaic (PV) net-metered system, as shown in Figure 2-1.[1]

In this example, a solar PV array on the roof (or a ground mounted system in a yard), produces a direct current (DC) voltage when the sun is out. The DC voltage then runs through an inverter which converts the DC electricity to alternating current (AC) electricity. The utility or the solar provider usually monitors the inverter in some way; often a separate circuit-breaker panel is installed between the solar PV system and the grid. Thus, in most systems, the AC voltage goes straight to the main utility breaker panel, which means it serves on-site loads first, and then if any electricity is available in excess of usage, the excess can flow onto the electric grid.

RELEVANT METERING TECHNOLOGIES

When the solar PV system is producing more electricity than the building is using—for example, perhaps when the occupants are away from home, few appliances are turned on, and the sun is shining—then excess solar-generated electricity (i.e., any amount

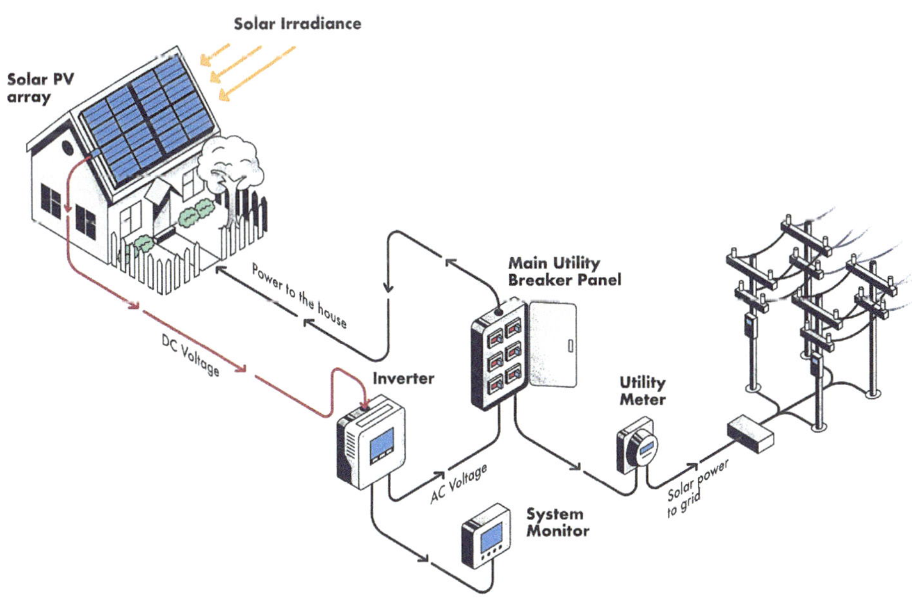

FIGURE 2-1 Technical configuration of a behind-the-meter rooftop solar system. The figure illustrates system operation when solar electricity generated exceeds total electricity load.

[1] Figure informed by Grafman, L., and J. Pearce. 2021. *To Catch the Sun*. Arcata, CA: Humboldt State University Press.

generated that is greater than the on-site demand at that moment) can be fed back into the grid. From a metering point of view, in older systems and under those circumstances, one could literally see the utility meter spinning backward; at the end of each billing period only the net amount of electricity consumed or produced was recorded by the utility. Under a traditional net metering arrangement, the customer would pay the retail rate for net electricity consumed (i.e., for consumption that exceeds production) or be credited at the same retail rate for the amount of production that exceeds consumption, which was allowed to feed back into the grid.

With a two-meter system, electricity production and consumption are recorded separately. The PV system may not be connected to the house's breaker panel and might feed all of its electricity onto the grid. Gross production and gross consumption are measured and can be paid or charged at different rates.

It should be noted that from a technical and engineering point of view, a PV system can also be used to power a specific load (or specific loads) in a building. For example, a PV system with a small battery backup may be used to power outdoor lighting and have no connection to the grid and thus is not net metered. In addition, PV systems can be used to power buildings or groups of buildings in off-grid installations that are not necessarily ever interconnected with the wider electricity grid.

Differing levels of sophistication in measuring, recording, and communicating a customer's electricity usage have had different implications for rate design and billing, which are relevant for net metering and its variants. For traditional net metering, where the meter records the net amount of a customer's withdrawals or consumption of electricity from the grid at the end of a billing period, after injections (DG production) and withdrawals take place over the course of the month, there was a straightforward alignment between net retail sales and standardized retail prices that did not vary by time of day or day of the week. As metering technology has evolved to enable more precise recording and measurement of withdrawals from the grid and injections to it, rate designs (such as time-of-use rates) can include different pricing of power for net withdrawals or net exports during different time periods, providing a means for managing a dynamic system with a number of distributed resources efficiently and reliably. This is important to keep in mind because the utility's cost to supply power varies by time of day and by seasons of the year.[2] Table 2-1 lists different types of metering technologies and their measurement capabilities, that are commonly used in the United States.

---

[2] For additional information see DOE Office of Electricity. n.d. "Recovery Act: Time Based Rate Programs." https://www.smartgrid.gov/recovery_act/time_based_rate_programs; Farugui, A., R. Hledik, and S. Sergici. 2019. "A Survey of Residential Time-of-Use (TOU) Rates." The Brattle Group, Inc. https://www.brattle.com/wp-content/uploads/2021/05/17904 a_survey_of_residential_time-of-use_tou_rates.pdf.

TABLE 2-1 Meter Options for Net Metering

Energy/Usage Meter	Measures the total kilowatt-hours of electricity delivered to the customer from the utility.
Production or Generation Meter	Measures how much electricity the customer produced (e.g., from a solar PV system).
Dual Meter	Consists of both a standard usage meter and a separate production/ generation meter.
Bi-directional Meter (also sometimes called a net meter)	Measures both electricity delivered to and electricity produced by the customer and sent to the utility. Can be an interval (readings are taken at defined times, like every hour) or cumulative. (Meter must record usage on the first and last day of each billing period to calculate net usage for that period.)
Smart Meter	Enables two-way communication between the utility and the meter, as well as digital connectivity allowing the use of Web-based home energy management tools and possibly integrating compatible home devices such as electric vehicles, air conditioning, and BTM DER, including BTM DG.
Variable Rate Meter	Variable rate meters allow utilities to charge customers different rates at different times of the day. Customers in some utility service areas are also able to opt-in to a variable rate electricity plan, where rates change by the hour or other time period in a day (rather than being a fixed per-kWh charge) and can fluctuate above or below the rate paid by flat rate customers.
Time-of-Use Meter	Measures total kilowatt-hours of electricity delivered to the customer from the utility and also has a time switch and multiple registers to record when the electricity is used, generally in either on- or off-peak hours. The simplest form of variable rate meter.
Demand Meter	Measures maximum kilowatts consumed in a set time period (usually 15 minutes). Used for utilities with electricity rate designs that include non-coincident and coincident peak demand charges.* Generally for commercial and industrial customers.

* Customers can be charged based on their individual peak, which may not coincide with system peak (i.e., on a non-coincident peak basis). Alternatively, customers can be charged based on their demand at the time of the system peak (i.e., on a coincident peak basis).

NOTE: BTM = behind the meter; DER = distributed energy resources; DG = distributed generation; kWh = kilowatt-hour.

It should be noted that not all these different metering technologies are functionally available in all areas of the United States, in part due to differing types of regulation and rate recovery associated with metering technologies. This variation in availability may present challenges for implementing more-complex rate designs and net metering variants or alternatives in particular locations.

BTM DG SYSTEM CONFIGURATIONS

In addition to the basic configurations of BTM DG with net metering described in Table 2-1, electricity storage can be added (see Figure 2-2). For such a solar + storage system, surplus generation can be used to charge the storage system, and then the stored generation can be discharged and fed into the grid at another time, with the exported power compensated according to the net metering or a net metering variant applicable to that customer. Solar + storage can also provide resilience for the electricity consumer in the case of a grid outage. Depending on the configuration of their interconnection, residences and commercial buildings with solar + storage can maintain electricity for themselves for some period of time during system outages, but BTM DG customers without storage would not be able to do so.[3]

In addition to feeding generation back into the grid, DG can be used by a customer to support demand-response capabilities. These can take a variety of forms, including agreements between the utility and customer for reduced rates if the customer agrees to use their DG to reduce their power demands during peak periods when requested to do so by the utility.[4] Energy storage systems charged by excess generation from BTM DG can also be used for demand response for load shaving during peak periods and load shifting[5] to off-peak periods or periods with excess generation. As mentioned, BTM DG and stored generation can provide resilience as well when there are system outages. Net metering policies can affect how storage is deployed and utilized, in conjunction with

[3] In some systems, the customer may have the ability to connect their BTM solar system to an internal load if there is a grid outage, enabling some level of more affordable resilience without a battery. Note that this configuration is prohibited in some areas.

[4] For more information see McDermott, T.E., K. McKenna, M. Heleno, B.A. Bhatti, M. Emmanuel, and S. Forrester. 2022. *Distribution System Research Roadmap*. Report No. PNNL-31550. Richland, WA: Pacific Northwest National Laboratory.

[5] For more information on load shifting see FERC. 2020a. "2020 Assessment of Demand Response and Advanced Metering." Staff report. https://cms.ferc.gov/sites/default/files/2020-12/2020%20Assessment%20of%20Demand%20Response%20and%20Advanced%20Metering_December%202020; and Potter, J., E. Stuart, and P. Cappers. 2018. *Barriers and Opportunities to Broader Adoption of Integrated Demand Side Management at Electric Utilities: A Scoping Study*. Berkeley, CA: Ernest Orlando Lawrence Berkeley National Laboratory. https://emp.lbl.gov/publications/barriers-and-opportunities-broader.

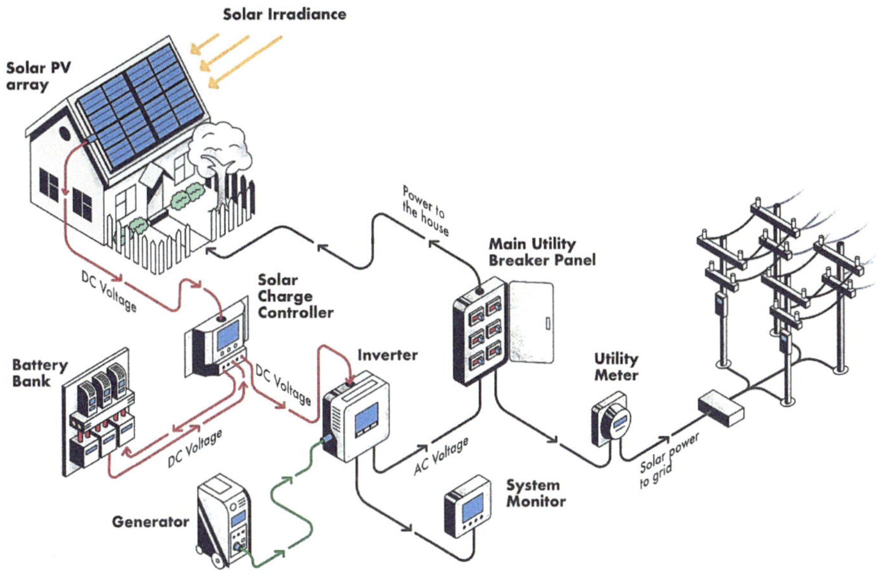

FIGURE 2-2 Technical configuration of a behind-the-meter rooftop solar system with battery storage.[a] The figure illustrates system operation at times when the electricity generation exceeds the demand in the building, but current flows in some of the wires reverse when the solar output is small.

[a] It should be noted that this figure illustrates a DC-coupled PV+storage system. Other examples of BTM PV+storage include AC-coupled PV+storage system. See the following for more information: Clean Energy Group. 2023. "Resilient Power Project." https://www.cleanegroup.org/publication-project/resilient-power-project; and DOE Office of Energy Efficiency & Renewable Energy. n.d. "Solar and Resilience Basics." https://www.energy.gov/eere/solar/solar-and-resilience-basics.

BTM DG, by providing compensation for electricity exported to the grid during peak demand periods (and at other times).[6]

The eventual emergence of newer DG and other DER configurations and the corresponding net metering compensation methods will depend on the development of new telecommunications, monitoring, networks of Internet of Things, power electronics, and other emerging technologies. These advances may also eventually allow for more opportunities for BTM DG, storage, demand response, and their increased adoption. There are many potential outcomes for how these systems can either take advantage of, or be limited by, net metering rules. Chapter 6 addresses these emerging technologies, the opportunities they provide, the challenges that need to be addressed, and the implications for pricing and increased DER adoption.

[6] Net metering policies may require that storage be charged with BTM DG, not grid power, to be included in net metering.

Background and History, Current Status, and Near-Term Future of Net Metering

INTRODUCTION

This chapter provides a review of net metering in the United States, from its beginnings in the early 1980s[1] through its widespread use by states and utilities in the 1990s and 2000s. Topics presented in this chapter include:

- Trends in adoption of net metering and related solar support policies
- Trends in adopting variants and alternatives to net metering in states, and their relative popularity
- Trends in deployment and costs for distributed solar generation and storage systems
- Trends and tensions in the relationship between electricity market structures, net metering policy adoption, and the introduction of solar-supporting policy mechanisms, as well as related developments and changes in energy and electricity system policy objectives
- Findings and recommendations

TRENDS IN THE ADOPTION OF NET METERING AND RELATED SOLAR SUPPORT POLICIES

The earliest state adopters of net metering did so for a number of reasons. According to a 1998 study by National Renewable Energy Laboratory (NREL) researchers:

[1] Eligible technologies and system sizes varied among the states, but typically provisions were included for small-scale photovoltaics and wind energy. Many states included other generating technologies, such as biomass, combined heat and power (CHP), small or micro-hydroelectric, and solar thermal electric. As reported in this chapter, 97 percent of all net metering participation has been for solar photovoltaic (PV) systems, with very small numbers of other technologies making up the remainder, mostly small wind and small or micro-hydroelectric systems. For state-by-state details of eligible technologies and other rules, see NC-CETA (North Carolina Clean Energy Technology Center). 2022a. "Database of State Incentives for Renewables & Efficiency® (DSIRE)." https://www.dsireusa.org.

> The main objective for states implementing net metering programs [was] to encourage private investment in renewable energy resources. Other goals include[d] stimulating local economic growth, diversifying energy resources, and improving the environment.[2]

Since then, other objectives have been added by states, including giving customers options for producing their own power and managing their energy bills, driving down the costs of renewables as the markets expand and technologies improve, and producing power with lower carbon emissions.[3] Net metering policies supported electricity customers' adoption of solar and other eligible behind-the-meter (BTM) resources and helped to satisfy these objectives.

Through the early 2010s, net metering was widely accepted as one of the key policies for supporting small residential and commercial distributed generation resources, most notably for solar photovoltaic (PV) systems. As shown in Table 3-1, 7 states initiated net metering programs in the 1980s, followed by 14 more states in the 1990s, and then another 21 states in the early 2000s.[4]

Net metering policies spread to many jurisdictions because net metering was a relatively easy to understand and administer tariff for serving early adopters of BTM distributed generation (DG) technologies. Almost all participating jurisdictions limited the total program size to small percentages of utility customers or utility load (initially not more than 1 percent in several states), such that any mismatch between benefits and costs would result in only de minimis shifting of benefits and costs between utility shareholders and customers, or between participating and non-participating customers.[5]

---

[2] Wan, Y.H., and H.J. Green. 1998. "Current Experience with Net Metering Programs." CP-500-24527. Golden, CO: National Renewable Energy Laboratory. Paper presented at WINDPOWER '98, Bakersfield, CA. https://www.nrel.gov/docs/legosti/old/24527.pdf.

[3] Many states have made formal commitments to reduce greenhouse gas emissions associated with energy production and use. NRRI (National Regulatory Research Institute). 2022b. "State Clean Energy Policy Tracker." https://www.naruc.org/nrri/nrri-activities/clean-energy-tracker.

[4] The term "state" is used here to refer to states or other regulatory jurisdictions, like the District of Columbia. Seven states started net metering programs in the 1980s: Massachusetts and Wisconsin (1982), Minnesota and New Hampshire (1983), Iowa (1984), Maine (1987), and Oklahoma (1988). See Stanton, T., and D. Phelan. 2013. "State and Utility Solar Energy Programs: Recommended Approaches for Growing Markets." NRRI 13-07. Silver Spring, MD: National Regulatory Research Institute. https://pubs.naruc.org/pub/FA86BBED-D69F-C1EC-D308-0BB2CDAF5E37.

[5] See Stanton, T. 2019. "Review of State Net Energy Metering and Successor Rate Designs." NRRI 19-01. Silver Spring, MD: National Regulatory Research Institute. https://pubs.naruc.org/pub/A107102C-92E5-776D-4114-9148841DE66B (hereafter Stanton 2019). Net metering was generally understood to be a "rough-justice" policy: "When net metering was just getting started, the small number of participating customers multiplied by the small quantity of energy each would deliver to the grid, meant that any error associated with under- or over-estimating the true value [of net metering credits] would be small."

TABLE 3-1 Timeline of Major State Actions on Net Metering Policy

1980s	1990s	2000s	2010	2011	2012	2013	2014	2015	2016	2017	2018	2019	2020	2021	2022
Cumulative number of jurisdictions with net metering policy in place at year end															
8[a]	21[a]	42[a]	43[a]	44	44	44	43	41	39	35	34	33	33	31	30
Expanding technologies eligible for net metering (shown in parentheses when adding provisions for including energy storage)															
		8 states							(HI)		(CA) (CO) (NY)	(AR) (MA) (SC)		(ME)	(MS)
Increasing size limits for individual systems (shown in parentheses for increasing size limit for total net metering program capacity)															
	MA	13 states	CA NJ VA WV	LA NV VA	IL IN MA NY	MN NY			MA		NJ SC	AR ME	VA	CO HI IL MA MD	(DE) MA MS

Changing carryover rules and/or credit rates for monthly net excess generation															
	MA **NH**	**14 states**	**VA** **WV**			**MN** **NV**				**NH** **NV**	**MA**			**KY**	**DE**
Replacing net metering with successor tariffs [with square brackets indicating initiating a first NEM-type tariff] and (in parentheses for court orders reversing net metering replacement tariff)															
							[MS][b]	**HI** **(NV)**	**AZ** **CA**	**HI** **(ME)** **MI[c]** **NY[d]** **UT**	**(KS)**	**LA**	**NY[d]** **UT**	**CT** **SC**	**IN**

[a] By year-end 2009, 41 states and the District of Columbia had adopted net metering policies. For the timeline of states initiating net metering, see K.M. Smith, C. Koski, and S. Siddiki, 2021, "Regulating Net Metering in the United States: A Landscape Overview of States' Net Metering Policies and Outcomes," *The Electricity Journal* 34(2):106901. For net metering policy changes prior to and shortly after 2010, see pp. 10 and 29 in T. Stanton, 2019, "Review of State Net Energy Metering and Successor Rate Designs," NRRI 19-01, Silver Spring, MD: National Regulatory Research Institute, https://pubs.naruc.org/pub/A107102C-92E5-776D-4114-9148841DE66B, and pp. 8–21 in T. Stanton and D. Phelan, 2013, "State and Utility Solar Energy Programs: Recommended Approaches for Growing Markets," NRRI 13-07, Silver Spring, MD: National Regulatory Research Institute, https://pubs.naruc.org/pub/FA86BBED-D69F-C1EC-D308-0BB2CDAF5E37.

[b] Mississippi calls its program net metering, but the provisions meet the definition of net-energy billing.

[c] The Michigan Public Service Commission issued an order approving a successor tariff design in 2017, but successor tariffs are being approved and implemented in individual utility rate cases following that decision.

[d] The New York Public Service Commission adopted a successor tariff for large commercial and community DG systems in 2017 and a successor tariff for other DG systems in 2020.

NOTE: See Appendix B for more details and links to official documents.

Although details might not have been explicitly addressed in authorizing legislation or in regulatory orders approving net metering tariffs, some industry participants believed net metering to be a temporary and participation-limited option. Many state programs were initiated in what were generally understood to be uneconomic or pre-economic markets, where solar PV installation costs were high compared to retail utility rates and returns on investment for participating customers were small, if any. Thus, rapid growth and expansive participation was not necessarily anticipated. In many jurisdictions, sunset clauses or trigger mechanisms calling for the review and resetting of policies were included in laws and rules, in case participation did reach specific limits in terms of capacity or numbers of participating customers. Even in jurisdictions with multiple types of taxpayer and ratepayer incentives, in the early stages of net metering programs only small numbers of potential consumers were adopting net metering systems.[6] As described in Chapter 1 of this report, net metering essentially involves customer generators reducing their utility bill at the retail rate for each kilowatt-hour of energy generated and used onsite. However, net metering programs differed, especially in terms of:

- Eligible generating technologies and the maximum size of eligible generators, usually related to the capacity needed to serve the customer's annual average energy usage
- Statewide program or utility system caps on participation usually expressed in percentage caps on participating capacity or numbers of customers
- Treatment of exported generation, including variations in the rate of compensation, if any, for that exported generation[7]

Gradually, solar PV technology improved, system costs declined, and financing options for distributed solar became more attractive to more customers. At the same time, net metering policies combined with multiple solar support mechanisms and financial incentives in many jurisdictions to further improve the customer economics for solar. As a result, larger numbers of customers began adopting BTM DG, particularly solar, and participating in net metering programs. Furthermore, with the declining costs of battery

[6] Taylor, M., J. McLaren, K. Cory, T. Davidovich, J. Sterling, and M. Makhyoun. 2015. "Value of Solar: Program Design and Implementation Considerations." NREL/TP-6A20-62361. Golden, CO: National Renewable Energy Laboratory. https://www.nrel.gov/docs/fy15osti/62361.pdf. Taylor et al. characterize markets for solar PV as non-economic when the lifecycle cost of PV energy is substantially higher than the price of utility-delivered energy, as pre-economic (or transitional) when the PV cost and grid price are converging, and as grid-competitive when PV cost is lower than grid price. As they note, combinations of net metering and other solar incentive and support policies in many jurisdictions gradually shifted markets in many utility service territories toward grid-competitive status, attracting increasing numbers of participating customers. See also Stanton 2019, pp. 4–7, note Stanton uses "uneconomic" instead of "non-economic."

[7] Smith, K.M., C. Koski, and S. Siddiki. 2021. "Regulating Net Metering in the United States: A Landscape Overview of States' Net Metering Policies and Outcomes." *The Electricity Journal* 34(2):106901.

storage, growing numbers of customers are pairing BTM DG with storage. Storage introduced new complexities to existing net metering policies, and several states (including Arkansas, California, Hawaii, Maine, Massachusetts, Mississippi, New York, and South Carolina) have clarified net metering rules to specify that paired systems are eligible, frequently with stipulations that energy delivered from batteries can receive net metering credits only if the batteries are charged from the qualifying BTM DG facility.

Net metering policy in the United States was frequently implemented at the same or nearly the same time as other state policies regarding clean and renewable energy financial support and as energy portfolio standard goals or requirements for some or all electric utility companies (and for competitive load serving entities in restructured states that allow retail competition). These are typically called renewable portfolio standards (RPS) or clean energy standards (CES). These policies established increasing quotas for electricity supplies from qualifying sources of generation and let markets and utility competitive bidding practices establish the prices for procuring and delivering those supplies. Qualifying technologies (including whether solar counted toward their goals), percentage goals, and target years differed widely among the states, but in the committee's judgment the practice most often directed utilities to produce or procure electricity that would add approximately 1 percent of new qualifying resources each year.[8] Figure 3-1 is a map summarizing the applicable portfolio policies in the states and territories. As shown in Figure 3-1, 15 jurisdictions are striving to achieve 100 percent of electricity production from qualifying resources by 2050 or sooner. A related policy in many U.S. states was to allow or to mandate that regulated utilities would offer green or renewable energy premium pricing programs, where customers could voluntarily participate in utility rate offerings with higher than the minimum and up to as much as 100 percent of supply from clean or renewable resources.[9]

---

[8] As experience was gained and costs fell as more net metering systems were installed, many states revisited their renewable or clean energy portfolio standards and increased the rates of additions-per-year. Gradually, many states enacted new laws and rules, increasing annual new clean energy capacity goals or mandates. In addition, many states have also made formal commitments to reduce greenhouse gas emissions associated with energy production and use. See NRRI. 2022b. "State Clean Energy Policy Tracker." https://www.naruc.org/nrri/nrri-activities/clean-energy-tracker.

By 2022, 41 states, the District of Columbia, and all 4 major U.S. territories had implemented renewable or clean energy goals or mandates, and 47 states, the District of Columbia, and all 4 territories had implemented net metering policies or had already enacted net metering alternatives. Both energy portfolio standards and net metering or net metering alternatives were in place in 35 states plus the District of Columbia and the 4 major territories. See NC-CETA. 2022a. "Database of State Incentives for Renewables & Efficiency® (DSIRE)." https://www.dsireusa.org.

[9] NREL. 2022b. "Energy Analysis, Voluntary Green Power Procurement." https://www.nrel.gov/analysis/green-power.html.

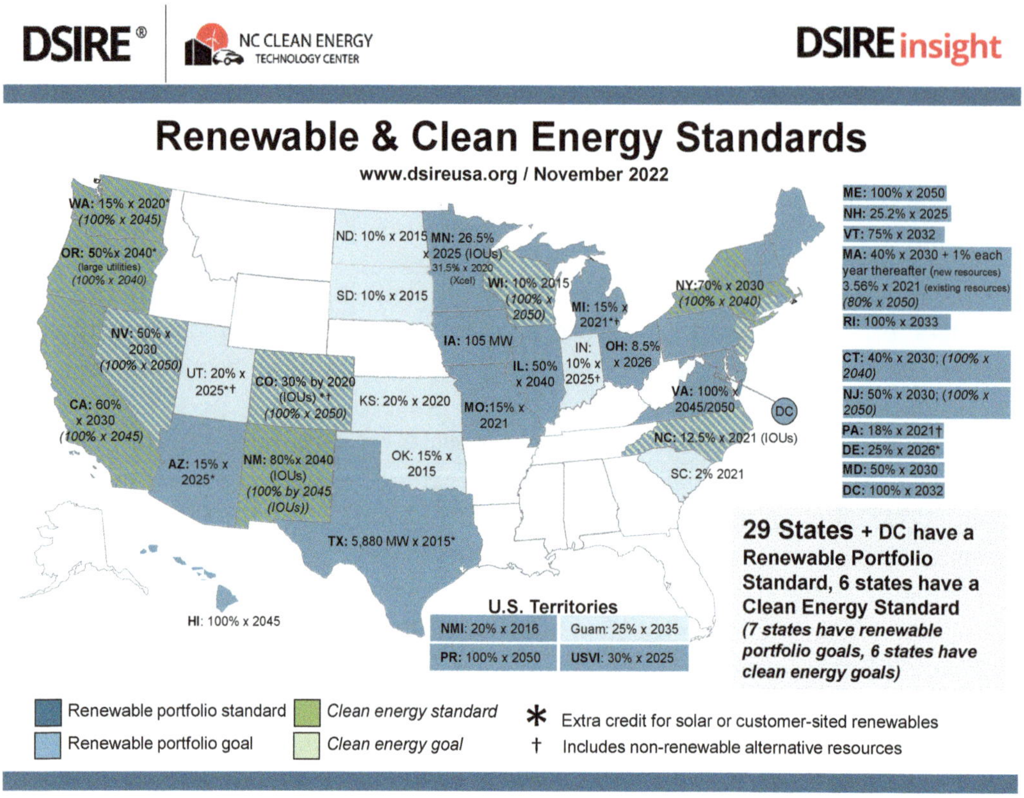

FIGURE 3-1 Renewable and clean energy standards detailed summary map (last updated November 2022).
SOURCE: NC-CETC (North Carolina Clean Energy Technology Center), 2022a, "Database of State Incentives for Renewables & Efficiency® (DSIRE)," https://www.dsireusa.org.

In specific states and utility service territories, rapid growth in customer-sited PV capacity increased pressure for modifying rate structures. Efforts have been made in many areas to adjust net metering variants and alternatives so that returns on investment for self-generation would not be excessive when considering multiple categories of benefits and costs associated with the self-generation, and the combinations of both utility rate treatment and the available federal, state, and sometimes also local solar support policies. In order to broaden participation by more consumers, many states added provisions for what was typically called aggregated or virtual net metering (VNM).[10]

[10] See the discussion of virtual net metering in Chapter 1 and Box 3-1.

TRENDS IN THE ADOPTION OF NET METERING VARIANTS AND ALTERNATIVES

The timeline presented in Table 3-1 illustrates this history of jurisdictions adopting net metering policies, followed by a gradual decline in traditional net metering as more jurisdictions modified their net metering policies and several opted for new kinds of DG tariffs. The number of states with net metering is shown declining, starting in the mid-2010s, as multiple jurisdictions adopted successor or replacement tariffs, as well as alternatives to ratemaking support for DG, sometimes particularly for solar or solar-plus storage.

By 2022, nearly every U.S. regulatory jurisdiction had considered at least some variants or alternatives to net metering.[11] These efforts arose in different ways: some state legislatures, such as California, directed regulators to transition to net metering variants and alternatives; utility regulators in other states, like Hawaii and New York, initiated regulatory proceedings to design variants; and utilities in some states, including Arizona, filed proposed net metering successor tariffs with their public utilities commissions. Many states had included in net metering enabling legislation sunset clauses or trigger mechanisms directing regulators to consider net metering variants, alternatives, and successors once participation levels reached pre-established program or utility system caps. By the end of 2021, as shown in Figure 3-2, reforms in net metering policy had already been implemented or were under active consideration in over half of all U.S. states, and several jurisdictions had already established and begun to implement net metering rate design variants, including several jurisdictions adopting what are often called net metering replacement tariffs.[12] Arizona, California, Hawaii, Indiana, Louisiana, Michigan, New York, South Carolina, and Utah had adopted variants to replace traditional net metering offerings. Other states, including Alabama, Mississippi, and South Dakota offer variants to traditional net metering, without previously having had in place a traditional net metering retail rate compensation structure. Georgia also fits into this latter category, except for a small net metering pilot program in 2020–2021 for customers of Georgia Power Company (offering net metering to 5,000 rooftop solar customers, up to 32 megawatts [MW] of newly installed total capacity).[13]

---

[11] See NC-CETC. 2022c. "The 50 States of Solar: Q4 2021 Quarterly Report and 2021 Annual Review." https://nccleantech.ncsu.edu/wp-content/uploads/2022/01/Q4-21-Solar-Exec-Summary-Final.pdf, pp. 7, 9, and 14.

[12] NC-CETC. 2022b. "The 50 States of Solar: 2021 Policy Review and Q4 2021 Quarterly Report." January. https://nccleantech.ncsu.edu/the-50-states-reports, p. 7.

[13] GPSC (Georgia Public Service Commission). 2020. "Order Approving the Renewable and Non-Renewable Resources Tariff (RNR-10) with Modifications." GPSC Docket No. 42516, Document Filing #181467. https://psc.ga.gov/search/facts-document/?documentId=181467.

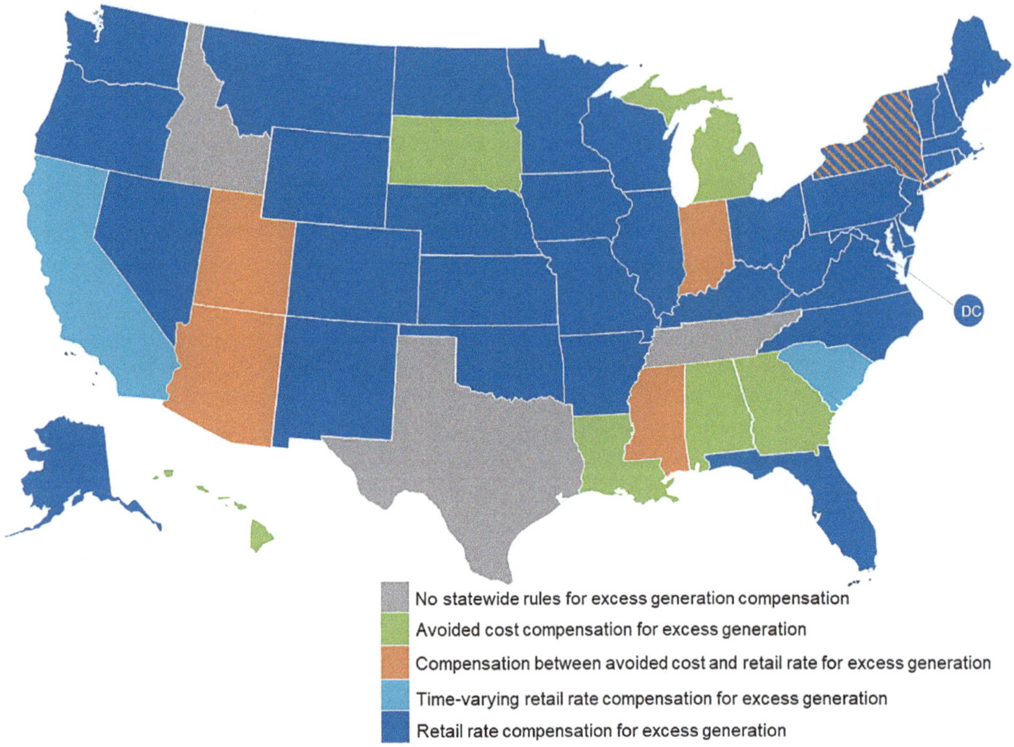

FIGURE 3-2 Credit rates used for excess generation in November 2022.
NOTES: Illinois is set to transition to compensation at the energy supply rate in 2025. Idaho and Texas do not have statewide rules for DG compensation, but several Idaho utilities and Texas competitive electric service providers do offer net metering for their participating customers.
SOURCE: NC-CETC (NC Clean Energy Technology Center), 2022d, "The 50 States of Solar: Q3 2022 Report," https://www.dsireinsight.com/publications.

Of states adopting net metering variants, the most common approach has been to implement a net billing structure, where netting intervals are shorter than the billing period and credit rates for exported energy differ from the retail rate. Most states that have approved a net billing structure are utilizing a netting interval of 1 hour or less; however, jurisdictions apply many variations of net billing credits. Inflow/outflow structures, where the kWh inflow and the kWh outflow are separately measured and priced, could be considered a form of net billing, where the netting is effectively performed on an "instantaneous" basis. Some states have elected to use a wholesale avoided cost rate as the compensation rate for excess generation, while others apply one or more adders or multipliers to the wholesale avoided cost rate. The latter group includes several states adopting their own value-of-solar (VOS) or value of distributed energy resources (VDER) rates.[14] For example:

- New York. For larger customer-generators, calculates credits using a VDER methodology
- Arizona is gradually reducing its credit rate by 10 percent per year until reaching the avoided cost, based on the recent contract cost of utility scale solar resources[15]
- California and South Carolina have adopted variants to traditional net metering, applying time-varying retail credit rates
- Massachusetts uses geographically variable retail credit rates that depend on costs associated with the specific distribution circuit where each DG project is interconnected[16]

Concerns about the potential for utility costs to be shifted from net metering customers to non–net metering customers continue as a major consideration in both legislative changes and regulatory proceedings exploring and implementing net metering variants and alternatives.[17] Separately from the net metering credit interval and rate, several states and utilities have attempted to address cost shifting issues primarily by assigning additional fees to customer-generators or in some cases to only those on net metering rates or those with BTM solar. Types of charges include demand charges, standby charges, fixed system use charges, and in a few cases minimum bills.[18] Although

[14] Personal communications with North Carolina Clean Energy Technology Center, November 2022. "DSIRE Insight Distributed Solar Policy Data. Value of Solar Studies" (spreadsheet); and NC-CETC. 2022b. "The 50 States of Solar: 2021 Policy Review and Q4 2021 Quarterly Report." January. https://nccleantech.ncsu.edu/the-50-states-reports, pp. 25–27.

In its review of distributed solar valuation studies, North Carolina Clean Energy Technology Center identifies 40 studies completed between 2005 and 2022. Studies have been completed in 26 different jurisdictions. In at least 16 of the jurisdictions, studies were undertaken by state public utility regulatory authorities or legislative working groups. However, not all the studies are used in establishing utility compensation rates for BTM DG. Minnesota and New York are basing compensation on the results of value-based studies for some BTM DG customers, but other states have not yet adjusted rates to reflect the findings of completed studies. Two other jurisdictions, Maryland and Missouri, are presently completing value-based studies. Table B-4, in Appendix B, lists the cost and benefit categories included in recently completed value-based studies.

[15] Arizona Corporation Commission. 2017. "Decision No. 75859." Docket No. E-00000J-14-0023. January 3. https://docket.images.azcc.gov/0000176114.pdf?i=1614371719161.

[16] For more information see https://www.mass.gov/guides/net-metering-guide#-2.-net-metering-credit-calculation-and-billing and https://www.mass.gov/files/220_cmr_18.00_final_12-1-17_1.pdf. This credit also only applied to net excess generation, so after net metering has already taken place.

[17] See further discussion of cost shifting in Chapters 4 and 5, including consideration that benefits of BTM DG supported by net metering can more than offset any costs.

[18] See Chapter 1 for discussion of ratemaking and definitions of terms.

several proposed fees for DG customers have been rejected by regulators or overturned by courts, certain investor-owned utilities in five states—Alabama, Arizona, New York, South Carolina, and Texas—have such fees in effect as of December 2022.[19]

TRENDS IN DEPLOYMENT AND COSTS FOR DISTRIBUTED SOLAR GENERATION AND STORAGE

Deployment Trends for Customer-Sited Generation and Storage

DER encompass a broad range of technologies—including DG, storage, energy efficiency, and demand response—that may be installed onsite or BTM by an electricity customer or connected directly to the utility's distribution network (in front of the meter). Net metering applies to customer-sited BTM generation and storage, with rules determining how participating customers are credited for electricity exported to the utility network. This section provides an overview of deployment trends for customer-sited generation and storage, as context for later discussions of the economic and equity implications of net metering.

Technology Trends

In aggregate, 72 gigawatts of customer-sited generation were installed in the United States as of year-end 2021, representing 5.6 percent of total U.S. electric generating capacity (albeit a smaller percentage of total electricity generation, given the lower capacity factor of solar compared to most other forms of generation).[20] As shown in Figure 3-3, that total consists almost entirely of PV and combined heat and power (CHP) systems, in roughly equal shares, with the remainder consisting of small quantities of wind, hydro, and other thermal (non-CHP) resources. As indicated in the figure, net metering

[19] NC-CETC. 2023. "The 50 States of Solar: 2022 Policy Review and Q4 2022 Quarterly Report." https://nccleantech.ncsu.edu/the-50-states-reports. For more information on each state see AL - https://www.alabamapower.com/content/dam/alabama-power/pdfs-docs/Rates/RGB.pdf; AZ - https://docs.tep.com/wp-content/uploads/714-TEP-Rider-14-RCP_P2.pdf; NY - https://www.nyserda.ny.gov/-/media/Project/Nyserda/Files/Programs/NY-Sun/Overview-of-Customer-Benefit-webinar-presentation.pdf; SC - https://p-cd.duke-energy.com/-/media/pdfs/for-your-home/rates/dep-sc/leaf-no-502-schedule-r-stou.pdf; TX - https://www.epelectric.com/files/html/Rates_and_Regulatory/TX%20Rates/Section%201%20-%20Sheet%2004.0%20-%20Schedule%2001%20Residential%20Service%20Rate.pdf.

[20] Percentage calculation is based on 72 GW (in alternating-current capacity) and a total 1273 GW (AC) nameplate capacity of U.S. utility-scale plus distributed electric generating plants through 2021, as reported by EIA. 2022c. "Electric Power Annual." November 7. https://www.eia.gov/electricity/annual.

and customer-sited solar are inextricably linked: almost all customer-sited solar is net-metered (92 percent), and almost all net-metered generation is solar (97 percent).[21] Although net metering policies in many states preceded the early years of this century, almost all customer-sited PV capacity has emerged since 2010, as shown in the yellow color in Figure 3-3.[22] That growth occurred as solar PV costs fell (as discussed later in this chapter), other supportive policies proliferated, and markets matured.[23]

Compared to customer-sited PV, customer-sited storage has been deployed at a much smaller scale to date, with just over 1 gigawatt installed in the United States as of the end of 2021 (see Figure 3-4). Virtually all that capacity has been built since 2017, consisting mostly of battery storage co-installed with solar PV. Nationally, about 10 percent of new residential PV installations and 5 percent of new commercial PV installations in 2021 were paired with storage.[24] This recent growth can be attributed to a combination of declining costs for lithium-ion battery storage technologies, policy incentives, and new sources of demand. One source of demand has been for backup power, most notably in California, where storage installations accelerated in the wake of wildfire-related Public Safety Power Shutoff (PSPS) events. Those trends are especially pronounced in the northern part of the state, where PSPS events have been most frequent.[25] A similar spike in paired PV-plus-storage occurred in Texas, following the extended power outages associated with winter storm Uri in February 2021.[26]

Importantly, demand for battery storage has also been driven by net metering reforms. Nowhere is this more evident than in Hawaii, where more than 80 percent of new PV installations are now paired with battery storage.[27] Since October 2015, when the State

[21] For additional data, see https://sgp.fas.org/crs/misc/R46010.pdf.

[22] Additional data sourced from https://irecusa.org/resources/annual-u-s-solar-market-trends-report.

[23] In contrast, commercial and industrial CHP is not net-metered and dates back much further in time, with most existing capacity installed between 1980 and 2005.

[24] Barbose, G.L., N. Darghouth, E. O'Shaughnessy, and S. Forrester. 2022. *Tracking the Sun: Pricing and Design Trends for Distributed Photovoltaic Systems in the United States*. Berkeley, CA: Lawrence Berkeley National Laboratory. https://emp.lbl.gov/tracking-the-sun.

[25] Barbose, G.L., S. Elmallah, and W. Gorman. 2021a. *Behind the Meter Solar + Storage: Market Data and Trends*. Berkeley, CA: Lawrence Berkeley National Laboratory. https://emp.lbl.gov/publications/behind-meter-solarstorage-market-data.

[26] Wood Mackenzie and Solar Energy Industries Association. 2022a. "US Solar Market Insight: 2021 Year-in-Review." https://www.woodmac.com/reports/power-markets-us-solar-market-insight-2021-year-in-review-150013339.

[27] Bai, J., D. Shuai, and E. Tian. 2022. "Solar PV Battery Installations in Honolulu: 2021 Update." Research and Economic Analysis Division, Department of Business, Economic Development and Tourism, State of Hawaii. https://files.hawaii.gov/dbedt/economic/data_reports/reports-studies/Solar_PV_Battery_Installation_2022May.pdf.

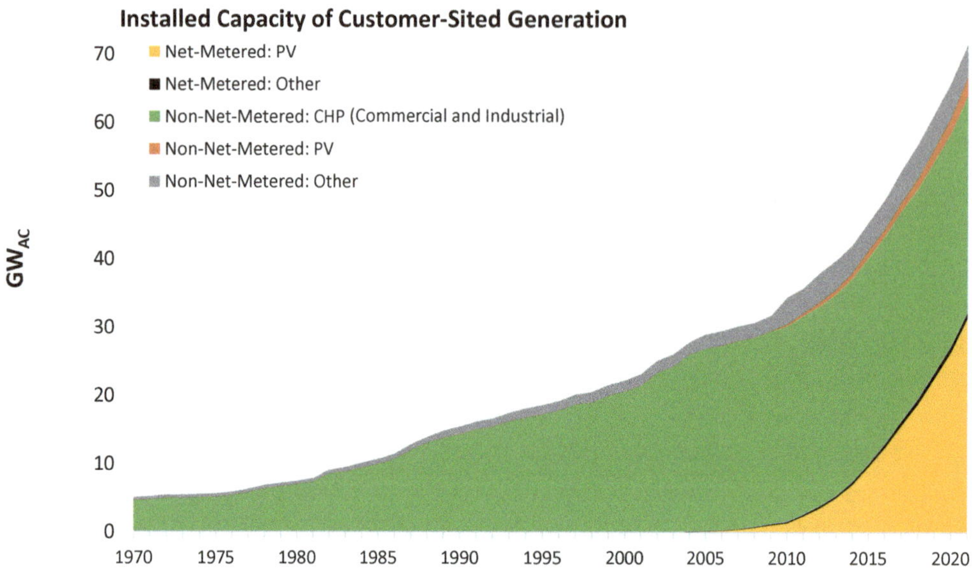

FIGURE 3-3 Historical growth of U.S. customer-sited generation.
NOTES: The figure aggregates data from several separate EIA datasets. Data for net-metered systems come from EIA Electric Power Annual, "Table 4.10. Net metering customers and capacity by technology type, by end use sector" (EIA 2022a). For the purpose of this figure, we exclude capacity associated with virtual net metering, which often consists of offsite projects. Data for small (<1 MW) non-net-metered systems are from EIA Electric Power Annual "Table 4.9.B. Total capacity of non-net-metered distributed generators by technology type and by sector" (EIA 2022a). For the purpose of this figure, we exclude capacity directly connected to the utility distribution network. Finally, data for large (≥1 MW) non-net-metered systems are from EIA Form-860 and consist of projects identified as commercial or industrial sectors (EIA 2022b). As such, the CHP totals exclude a sizable amount of capacity owned by independent power producers.
SOURCES: Data from EIA, 2022a, "Annual Electric Power Industry Report, Form EIA-861 Detailed Data Files," November 18. https://www.eia.gov/electricity/data/eia861; EIA, 2022b, "Annual Energy Outlook," March 3, https://www.eia.gov/outlooks/aeo/ index.php.

Public Utilities Commission limited the number of new solar PV customers to send excess power to the electric grid for credit, battery-connected PV system then become a more attractive option. As that transition has progressed, customers have increasingly paired new solar installations with storage, which they use to reduce exports and maintain full retail rate offsets for as much of their PV generation as possible. As a result, revenue impacts to the utility are similar to what would have occurred under traditional net metering. (See Chapter 4 for further discussion.)

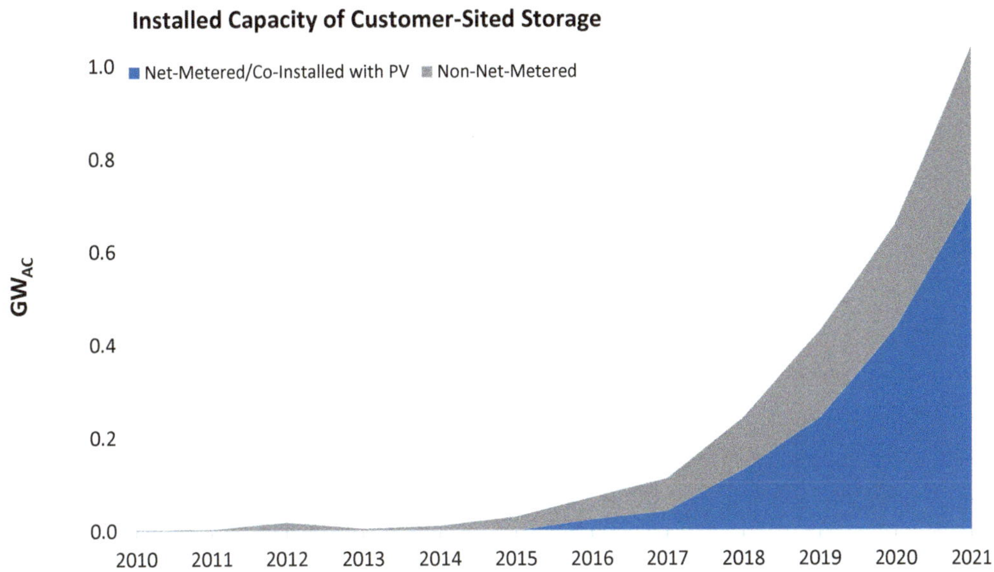

FIGURE 3-4 Historical growth of U.S. customer-sited storage.
NOTES: The figure aggregates data from several separate EIA datasets. Data for net-metered storage co-installed with PV are from EIA Electric Power Annual, "Table 4.10. Net metering customers and capacity by technology type, by end use sector."[a] Data for small (<1 MW) non-net-metered storage are from EIA Electric Power Annual, "Table 4.9.B. Total capacity of non-net-metered distributed generators by technology type and by sector."[a] For the purpose of this figure, we exclude capacity directly connected to the utility distribution network. Finally, data for large (≥1 MW) non-net-metered storage are from EIA Form-860 and consist of projects identified as commercial or industrial sector.[b]

[a] EIA. 2022b. "Annual Energy Outlook." March 3. https://www.eia.gov/outlooks/aeo/index.php.

[b] EIA. 2022e. "Form EIA-860 Detailed Data with Previous Form Data (EIA-860A/860B)." June 2. https://www.eia.gov/electricity/data/eia860.

SOURCES: Data from EIA, 2022b, "Annual Energy Outlook," March 3, https://www.eia.gov/outlooks/aeo/index.php; EIA, 2022e, "Form EIA-860 Detailed Data with Previous Form Data (EIA-860A/860B)," June 2, https://www.eia.gov/electricity/data/eia860.

Sectoral and Geographic Trends for Customer-Sited PV

The total installed capacity of customer-sited PV in the United States is split roughly 60/40 between the residential and commercial and industrial (C&I) sectors, but growth in recent years has skewed more heavily toward the residential market. In 2021, 75 percent of annual capacity additions were installed by residential customers (see Figure 3-5). In general, C&I customers tend to be more price-sensitive, and the C&I market has seen more volatility as state-level policies and incentives ebb and flow, resulting in relatively flat growth since 2016. These numbers do not include the relatively small but burgeoning community solar sector (see Box 3-1).

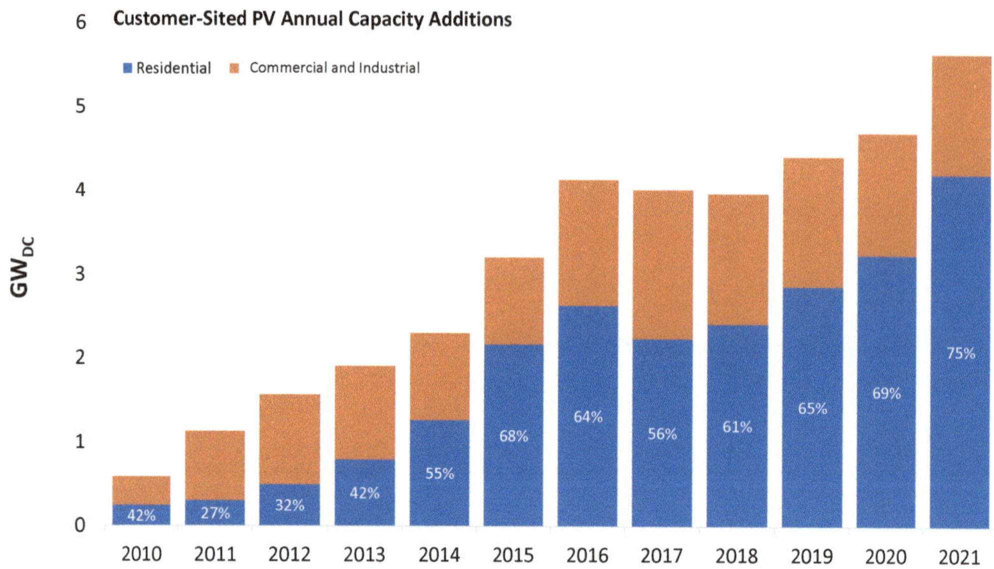

FIGURE 3-5 Sectoral trends for U.S. customer-sited PV.
NOTES: Data for the commercial and industrial sector exclude community solar. Note that Wood Mackenzie and Solar Energy Industries Association reports capacity in DC units; for that reason, and others, the values cannot be compared directly against the EIA-based totals in previous figures.
SOURCE: Data from Wood Mackenzie and Solar Energy Industries Association, 2022a, "US Solar Market Insight: 2021 Year-in-Review," https://www.woodmac.com/reports/power-markets-us-solar-market-insight-2021-year-in-review-150013339.

At the national level, BTM DG market penetration levels are still in the low single-digits. In the residential sector, roughly 3.2 million customer-sited PV systems were installed by the end of 2021, representing 2.3 percent of all U.S. residential electricity customers and 4.6 percent of all owner-occupied, detached single-family homes, which constitute most adopters.[28] Within the C&I sector, roughly 120,000 systems were installed through 2021, equivalent to 0.6 percent of all U.S. C&I electricity customers. C&I PV adopters come

[28] Barbose, G.L., S. Forrester, E. O'Shaughnessy, and N.R. Darghouth. 2021b. *Residential Solar-Adopter Income and Demographic Trends: 2021 Update*. Berkeley, CA: Lawrence Berkeley National Laboratory. https://emp.lbl.gov/publications/residential-solar-adopter-income-and; Wood Mackenzie and Solar Energy Industries Association. 2022a. "US Solar Market Insight: 2021 Year-in-Review." https://www.woodmac.com/reports/power-markets-us-solar-market-insight-2021-year-in-review-150013339. The 4.6 percent number is derived by dividing 3.2 million by the total number of owner-occupied detached single-family homes, which is from the 2021 American Community Survey Table B25032, available here: https://data.census.gov/table?t=Owner/Renter+(Tenure)&d=ACS+1-Year+Estimates+Detailed+Tables&tid=ACSDT1Y2021.B25032.

BOX 3-1
COMMUNITY SOLAR DEPLOYMENT TRENDS

Community solar refers to a particular business model, where multiple individual customers purchase shares of, or subscribe to, an offsite solar project and are compensated for their portion of the project's output via credits on their utility bills. Another variation for multi-family buildings, is one where tenants are invited to purchase shares or subscribe to an onsite system. Bill credits provided to subscribers are often calculated in the same manner as net metering—as a one-for-one reduction in billed consumption. This arrangement is sometimes referred to as "virtual net metering."

Roughly 4 GW of community solar capacity were installed in the United States through year-end 2021, with 1 GW installed in 2021 alone, approaching the size of the customer-sited C&I market (Burton and Kaifeng 2022; Wood Mackenzie and Solar Energy Industries 2022a). The Department of Energy (DOE) reports that 22 states and the District of Columbia had legislative provisions enabling or requiring community solar by year-end 2021. In 2021, DOE announced a target for its National Community Solar Partnership, for "community solar systems to power the equivalent of five million households by 2025 and create $1 billion in energy bill savings" (DOE 2021).

In 17 states, community solar legislation includes provisions for low-income participation (DOE, 2023). Some state government incentives provide additional funding for projects that include low-income customers among subscribers. Some states have mandates, called carve-outs, that require a certain percentage of a community solar project or program participation by low-income subscribers or organizations that provide services to low-income clientele; however, by the end of 2021, such carve-outs accounted for just 1 percent of all operational community solar capacity (Chan et al. 2022).

References

Burton, R., and X. Kaifeng. 2022. "Community Solar Resource Database." Golden, CO: National Renewable Energy Laboratory. https://data.nrel.gov/submissions/201.

Chan, G., J. Heeter, and K. Xu. 2022. "Sharing the Sun Community Solar Project Data (December 2021)." Golden, CO: National Renewable Energy Laboratory. https://data.nrel.gov/submissions/185.

DOE (Department of Energy). 2021. "DOE Sets 2025 Community Solar Target to Power 5 Million Homes." https://www.energy.gov/articles/doe-sets-2025-community-solar-target-power-5-million-homes.

DOE. 2023. "Community Solar Market Trends." https://www.energy.gov/communitysolar/community-solar-market-trends.

Heeter, J., K. Xu, and G. Chan. 2021. *Sharing the Sun: Community Solar Deployment, Subscription Savings, and Energy Burden Reduction*. NREL/PR-6A20-80246. Golden, CO: National Renewable Energy Laboratory. https://www.nrel.gov/docs/fy21osti/80246.pdf.

Wood Mackenzie and Solar Energy Industries Association. 2022a. "US Solar Market Insight: 2021 Year-in-Review." https://www.woodmac.com/reports/power-markets-us-solar-market-insight-2021-year-in-review-150013339.

from all industry sectors, albeit with some concentration among schools, warehouses, and big-box retail stores.[29]

Penetration levels vary dramatically across states and regions, reflecting policy conditions, electricity prices, and solar insolation levels. For the residential market, participation levels are highest in Hawaii, where 20 percent of residential electricity customers had installed PV by the end of 2021, followed by California at 10 percent (see Figure 3-6 for percentages of participation in other states). Other high-penetration states are primarily in the desert southwest, with abundant sunshine, and the northeast, where electricity prices are relatively high and additional financial incentives are available (often in the form of renewable energy certificates used for RPS compliance).[30] Two states (Arizona and Nevada) had reached the 5 to 10 percent penetration threshold, while 7 other states—in either the southwest or northeast—and the District of Columbia surpassed 3 percent penetration. Ten more states spanning a somewhat more diverse geography, including several states in the northwest and southeast, had penetration rates of 1–3 percent. The remaining 29 states all had PV penetration levels less than 1 percent of residential customers as of year-end 2021.

Cost Trends for Distributed PV and Storage

The impacts of net metering on DG adoption, and their implications for future net metering policies, must be viewed in the context of a broad set of market conditions and deployment drivers for DG. Chief among those drivers in the committee's judgment has been declining costs, particularly for rooftop PV, and to a lesser extent for distributed storage as well.

As shown in Figure 3-7, residential solar PV system costs have fallen by 69 percent ($8.50 per Watt) over the past two decades, while large commercial system costs fell by 77

[29] Hoen, B., J. Rand, and S. Elmallah. 2019. *Commercial PV Property Characterization: An Analysis of Solar Deployment Trends in Commercial Real Estate*. Berkeley, CA: Lawrence Berkeley National Laboratory. https://emp.lbl.gov/publications/commercial-pv-property. Wood Mackenzie and Solar Energy Industries Association. 2022a. "US Solar Market Insight: 2021 Year-in-Review." The 0.6 percent number is derived by dividing 120,000 by the total number of U.S. C&I electricity customers (from EIA Electric Power Annual, Table 2.1).

[30] Barbose, G. 2021. *U.S. Renewables Portfolio Standards 2021 Status Update: Early Release*. Berkeley, CA: Lawrence Berkeley National Laboratory. https://emp.lbl.gov/publications/us-renewables-portfolio-standards-3.

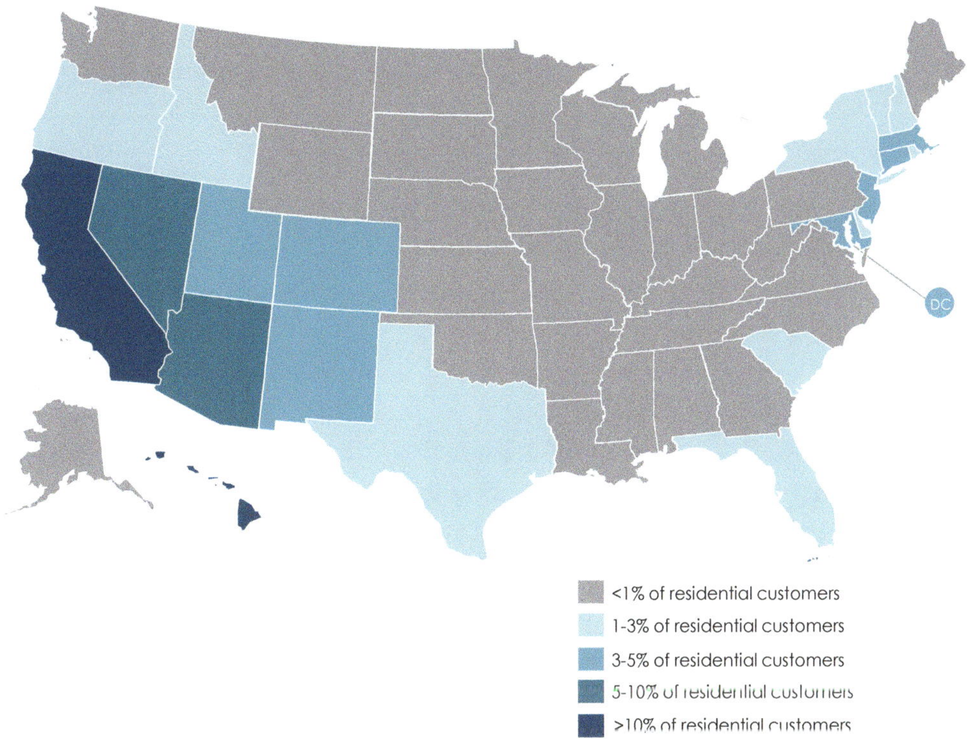

FIGURE 3-6 2021 statewide residential PV penetration.
NOTES: Penetration refers to the percentage of residential customers with distributed PV systems. As noted earlier, 92 percent of distributed PV is net-metered, and 97 percent of all net metered capacity is PV; thus, the chart effectively shows the penetration of all net-metered DG by residential customers.
SOURCES: Values are calculated from statewide residential PV system counts (Wood Mackenzie 2022a) and from the total number of residential electricity customers in each state (EIA 2022c).

percent.[31] That drop in system level costs has been driven to a significant degree by the cost of PV modules, which fell by $4.20/W over the same period while inverter prices fell by $0.90/W over the past two decades, also contributing to total PV system cost reductions.[32] The remainder of system-level cost declines are associated primarily with the various business process or "soft" costs, including installation labor, customer acquisition

[31] Barbose, G., N. Darghouth, E. O'Shaughnessy, and S. Forrester. 2022. *Tracking the Sun: Pricing and Design Trends for Distributed Photovoltaic Systems in the United States*. Berkeley, CA: Lawrence Berkeley National Laboratory. https://emp.lbl.gov/tracking-the-sun.

[32] Most of the decline in module costs occurred during the period from 2008–2012 when global annual PV module manufacturing capacity expanded more than sevenfold, from roughly 8 to 60 GW. (Bloomberg New Energy Finance [BNEF], global PV module manufacturing capacity.)

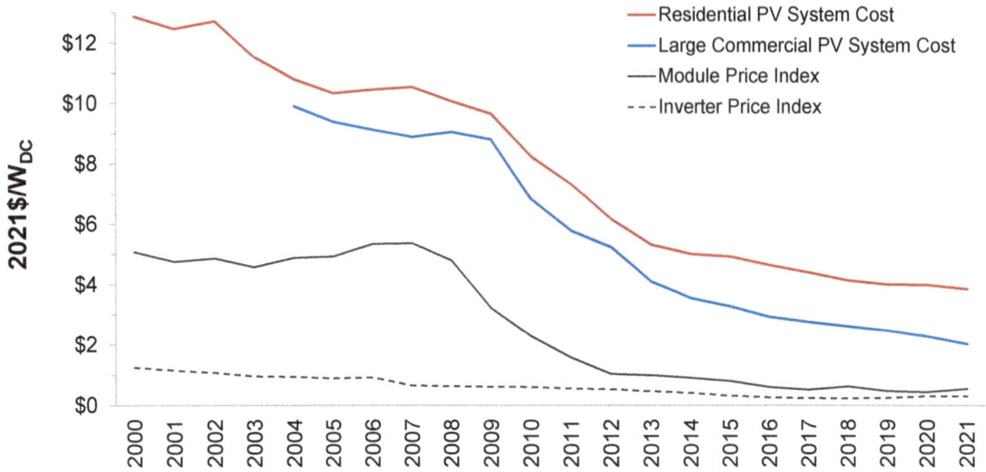

FIGURE 3-7 Cost trends for customer-sited PV in the United States.
NOTE: Non-residential PV system cost data are based on customer-sited systems larger than 100 kW; historical data for those systems are available only as far back as 2004.
SOURCE: Data from G. Barbose, N. Darghouth, E. O'Shaughnessy, and S. Forrester, 2022, *Tracking the Sun: Pricing and Design Trends for Distributed Photovoltaic Systems in the United States*, Berkeley, CA: Lawrence Berkeley National Laboratory, https://emp.lbl.gov/tracking-the-sun.

costs, permitting and inspection costs, as well as installer profit. Those soft costs now constitute most of the total cost of customer-sited PV systems.[33]

Distributed storage costs have also fallen, albeit over a much shorter and more recent timeframe than for rooftop PV. As shown in Figure 3-8, residential storage costs decreased by 23 percent ($0.80/W) since 2014, while commercial storage costs decreased by 37 percent ($0.90/W).[34] Much of the decline in the total cost of energy storage systems is due to the drop in the cost of lithium-ion battery packs, as global production has ramped up.

TRENDS AND TENSIONS IN NET METERING AND ASSOCIATED SUPPORTING POLICIES

Discussions about net metering and broader policies addressing DG and other DER are occurring within a multifaceted policy environment, where many pressing, interrelated

[33] Soft costs are the non-hardware costs associated with energy supply systems. They include, for example: "design, siting, permitting, installation, interconnection, and financing. They also include the sales, general, and administrative expenses solar companies incur for customer acquisition, workforce training and certification, supply chain and inventory control, and operating overhead." See EERE. 2022b. "Soft Costs." https://www.energy.gov/eere/solar/soft-costs.

[34] Data are sourced from Wood Mackenzie. 2022. "U.S. Energy Storage Market Monitor."

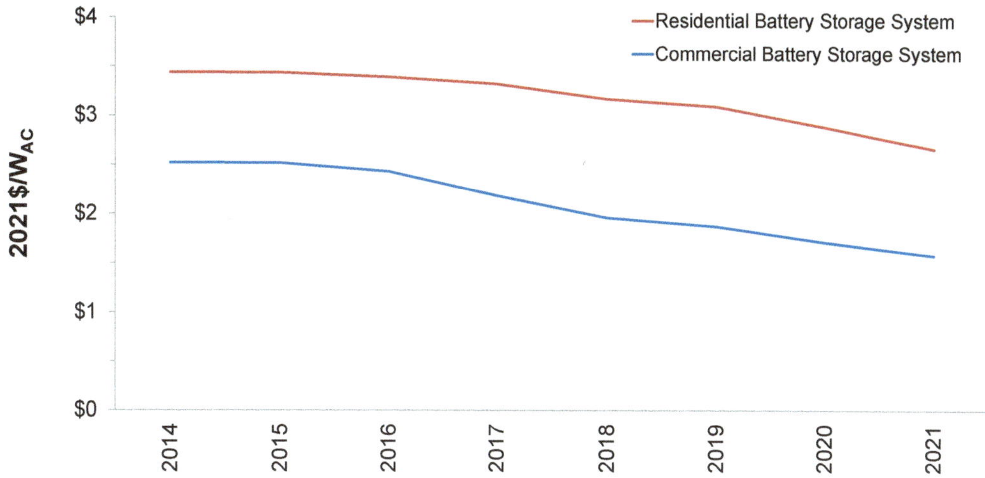

FIGURE 3-8 Cost trends for distributed battery storage in the United States.
NOTE: Costs are based on 2-hour duration lithium-ion batteries.
SOURCE: Data from Wood Mackenzie, 2022, "U.S. Energy Storage Market Monitor."

challenges are facing the public utility industry, its regulators, policymakers, and customers. The primary challenges and trends include:

- The need for major, rapid changes in energy supplies and demands to meet formal climate-change commitments made by government agencies, utilities, and many utility customers.[35]
- The need for major investments in the transmission and distribution system to replace aging assets, help achieve new policy goals (e.g., interconnecting DG and remote grid scale renewables), and meet changing customer expectations and demands, including those related to the adoption of BTM DER.[36]
- The need to address grid operational and stability challenges arising as a result of the increasing deployment of variable-output sources of generation, such as wind and solar, both at the distribution and transmission levels.[37]

---

[35] See Fox-Penner, P. 2020. *Power After Carbon: Building a Clean, Resilient Grid.* Cambridge, MA: Harvard University Press. https://www.hup.harvard.edu/catalog.php?isbn=9780674241077, pp. 3, 7–8, 22–27 and Chapter 6; and NRRI. 2022b. "State Clean Energy Policy Tracker." https://www.naruc.org/nrri/nrri-activities/clean-energy-tracker.

[36] The Department of Energy is presently engaged in a National Transmission Planning Study. See DOE Grid Deployment Office. 2022. "National Transmission Planning Study." https://www.energy.gov/gdo/national-transmission-planning-study.

[37] Impram, S., S.V. Nese, and B. Oral. 2020. "Challenges of Renewable Energy Penetration on Power System Flexibility: A Survey." *Energy Strategy Reviews* 31:100539.

- The opportunity to address these grid challenges through new and emerging technologies for infrastructure modernization and integration of distributed clean energy resources and energy storage.[38]
- Changing utility rate designs that better reflect underlying costs and influence consumer behaviors, including decisions to adopt DG and other DER.[39]
- Increasing explicitly stated directives from governors and legislators for state energy regulatory commissions to address equity, environmental justice, and closely related values (see Chapter 5 for further discussion).
- Changing drivers of electricity consumption—energy efficiency reducing usage on the one hand and electrification of transportation, buildings, and industry increasing it on the other.

The broader context also includes growing interest in utility business models and regulatory innovation, particularly as new and emerging technologies are deployed (see Chapter 7). This context also includes increasing challenges associated with the siting and community acceptance of all energy-related infrastructure.[40] While all of these are important issues, thorough treatment of them is beyond the scope of this report. What is most important for readers to understand is that existing net metering policies interact with these challenges and trends, and those interactions simultaneously help shape the environment in which policymakers are considering net metering and its variants and alternatives.

Some publicly available data can shed light on some of these drivers, including information collected at the utility-specific and jurisdiction levels by the Department of Energy's Energy Information Administration[41] and by North Carolina State University's NC Clean

[38] DOE Grid Modernization Initiative. 2020. "Grid Modernization Updated GMI Strategy 2020." https://www.energy.gov/sites/prod/files/2021/02/f82/GMI_Strategy_FINAL%20as%20of%201.20.21.pdf; and EIA. 2021b. "Solar Generation Was 3% of U.S. Electricity in 2020, But We Project It Will Be 20% by 2050." https://www.eia.gov/todayinenergy/detail.php?id=50357.

[39] The Behavior, Energy & Climate Change Conference (BECC) focuses on energy consumers and rate designs intended to change usage patterns. See BECC (Behavior, Energy and Climate Change). 2022. "About BECC." https://beccconference.org/about-becc. For performance-based regulation, see Littell, D., C. Kadoch, P. Baker, et al. 2017. *Next-Generation Performance-Based Regulation: Emphasizing Utility Performance to Unleash Power Sector Innovation*. NREL/TP-6A50-68512. Golden, CO: National Renewable Energy Laboratory; Sappington, D.E.M., and D.L. Weisman. 2021. "Designing Performance-Based Regulation to Enhance Industry Performance and Consumer Welfare." *Electricity Journal* 34(2):106902; Weisman, D. 2018. "A Report on the Theory and Practice of Performance-Based Regulation." SSRN. https://ssrn.com/abstract=3765691.

[40] Wilkinson, F. 2022. "How Copake, New York, Fought and Then Accepted a Solar Farm." Bloomberg, March 20. https://www.bloomberg.com/opinion/articles/2022-03-20/how-copake-new-york-fought-and-then-accepted-a-solar-farm.

[41] EIA. 2022d. "Electricity: Form EIA-861M (formerly EIA-826) Detailed Data." https://www.eia.gov/electricity/data/eia861m.

Energy Technology Center.[42] Additional information should be collected and made available to the public in order for researchers, analysts, utilities, policymakers, and others to more fully understand the drivers and outcomes of BTM DG and storage deployment, net metering and net metering variants adoption, electricity production, and interactions with the utility grid.[43]

INTERNATIONAL EXPERIENCES WITH DG-PV COMPENSATION POLICIES

Other countries, notably many countries in Europe, and a few in Asia, as well as Australia and Canada, used a different primary renewable and solar energy support policy, known as a feed-in-tariff (FIT; described as a net metering variant in Chapter 1).[44] With this approach, regulators and administrators establish long-term contract prices for energy from qualifying energy sources that are intended to provide reasonably attractive rates of return on investment. The FIT price is set high enough to attract investing, and then the quantity adjusts in response to market interest. The general objective for a FIT is to "provide a stable investment framework, thus lowering the investment costs and the costs for the final consumers."[45]

By the end of 2008, 50 countries had implemented feed-in tariffs, either nationwide or in at least some specific jurisdictions, and by 2010 FITs had spread to 63

[42] The NC Clean Energy Technology Center Hosts the Database of State Incentives for Renewables & Efficiency® (DSIRE). See NC-CETA. 2022a. "Database of State Incentives for Renewables & Efficiency® (DSIRE)." https://www.dsireusa.org.

[43] For example, utilities typically do not provide information about (and in some cases may not even have their own visibility into) customers' exports of power into the local distribution grid. More granular reporting of such information about individual BTM systems' exports to the grid (e.g., the timing and quantities of exports) would be helpful for understanding the impacts of net metering and DG on utilities and on both participating and non-participating utility customers.

[44] UNEP (United Nations Environment Programme). 2012. "Feed-In Tariffs as a Policy Instrument for Promoting Renewable Energies and Green Economies in Developing Countries." https://unfccc.int/files/documentation/submissions_from_parties/adp/application/pdf/unep_us___ws2.pdf, UNEP states. "The most prevalent national renewable energy policy in the world is the feed-in tariff (FIT). As of early 2011, 50 countries had some form of FIT in place, with more than half of these being in developing countries" (p. vi). See also Organisation for Economic Co-operation and Development (OECD). 2022. "Renewable Energy Feed-In Tariffs." https://stats.oecd.org/Index.aspx?DataSetCode=RE_FIT.

[45] Mendonça, M., and D. Jacobs. 2009. "Feed-In Tariffs Go Global: Policy in Practice." Renewable Energy World. https://www.renewableenergyworld.com/baseload/feed-in-tariffs-go-global-policy-in-practice. See also Glemarec, Y., W. Rickerson, and O. Waissbein. 2012. *Transforming On-Grid Renewable Energy Markets. A Review of UNDP-GEF Support for Feed-In Tariffs and Related Price and Market-Access Instruments.* Washington, DC: Department of Energy, Office of Scientific and Technical Information. https://www.osti.gov/biblio/22090461, p. 6.

jurisdictions.[46] Researchers observed that FITs succeeded in opening up opportunities for solar PV and wind generation in many countries.[47]

Nevertheless, observers cautioned that FITs require substantial "supervision and control,"[48] and that poorly designed and implemented FITs could "backfire" and even "stunt" the industry,[49] resulting in "unintended and counterproductive outcomes"[50] including potential boom and bust cycles. Experiences in Germany and Spain demonstrated problems with FITs: changes in market conditions in Germany resulted in substantial added costs for utility customers; in Spain, FIT policies were associated with a noteworthy boom and bust cycle.[51] It appears that the Spanish FIT was overly generous in compensating solar PV from 2005 through 2008, resulting in internal rates of returns to investors are estimated to have been between 10 to 15 percent, as opposed to the targeted rate of 5 to 9 percent. Then, as shown in Figure 3-9, major revisions to the FIT policy stopped new PV installations entirely in 2009 PV returned to something less than 2007 levels for the next few years.

In order to reduce costs while continuing growth, Germany shifted toward a policy combining FITs with auctions, intending to introduce more competition into markets for renewable energy. Germany first introduced auctions in 2014, and then shifted entirely to an auction-based process beginning in 2016.[52]

---

[46] Patel, S., and G. Reitenbach. 2010. "The Feed-In Tariff Factor." Power, September 1. https://www.powermag.com/the-feed-in-tariff-factor; Taylor, M., J. McLaren, K. Cory, T. Davidovich, J. Sterling, and M. Makhyoun. 2015. "Value of Solar: Program Design and Implementation Considerations." NREL/TP-6A20-62361. Golden, CO: National Renewable Energy Laboratory. https://www.nrel.gov/docs/fy15osti/62361.pdf. See also Zachary, J. 2010. "Feed-In Tariffs Around the World." Power, September 1. https://www.powermag.com/feed-in-tariffs-around-the-world.

[47] DeShazo, J.R., and R. Matulka. 2009. "Best Practices for Implementing a Feed-In Tariff Program." Los Angeles Business Council and UCLA Luskin Center for Innovation. https://innovation.luskin.ucla.edu/wp-content/uploads/2019/03/Best_Practices_for_Implementing_a_Feed-in_Tariff_Program.pdf.

[48] Barnea, G., C. Hagemann, and S. Wurster. 2022. "Policy Instruments Matter: Support Schemes for Renewable Energy Capacity in Worldwide Comparison." *Energy Policy* 168.

[49] Aflaki, S., and A. Masini. 2020. "Optimizing Feed-In Tariffs to Boost Renewable Energy Production." Jouy en Josas, Paris: HEC Paris. https://www.hec.edu/en/knowledge/articles/optimizing-feed-tariffs-boost-renewable-energy-production.

[50] DeShazo, J.R., and R. Matulka. 2009. "Best Practices for Implementing a Feed-In Tariff Program." Los Angeles Business Council and UCLA Luskin Center for Innovation. https://innovation.luskin.ucla.edu/wp-content/uploads/2019/03/Best_Practices_for_Implementing_a_Feed-in_Tariff_Program.pdf.

[51] del Rio, P., and P. Mir-Artiques. 2014. *A Cautionary Tale: Spain's Solar PV Investment Bubble*. Winnipeg, Manitoba, Canada: International Institute for Sustainable Development. https://www.iisd.org/gsi/sites/default/files/rens_ct_spain.pdf.

[52] Recent Facts About Photovoltaics in Germany, Harry Wirth, Fraunhofer ISE, download from https://www.ise.fraunhofer.de/en/publications/studies/recent-facts-about-pv-in-germany.html, version of May 15, 2021. For further information see https://www.iea.org/reports/germany-2020.

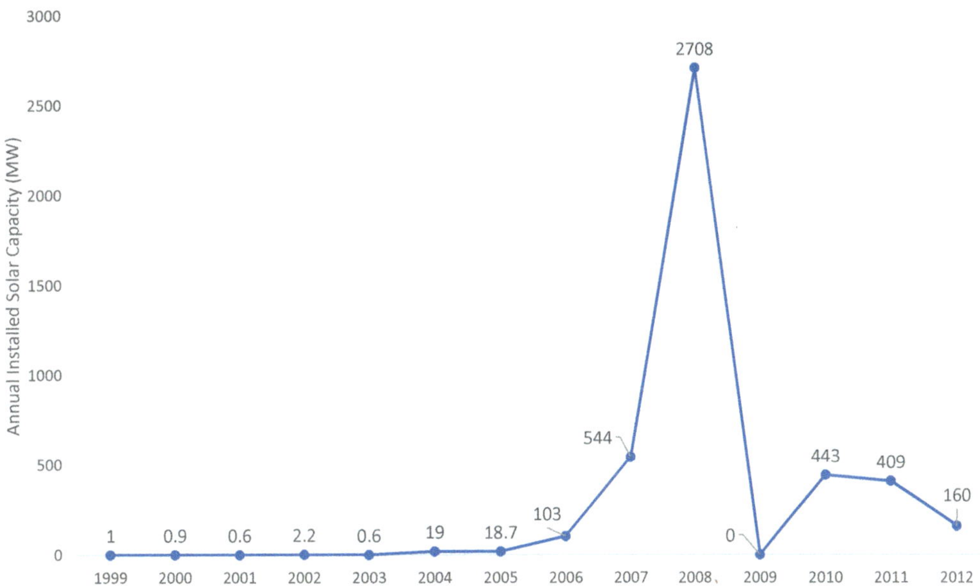

FIGURE 3-9 Annual installed PV capacity in Spain (in MW), 1999–2012.
SOURCE: P. del Rio and P. Mir-Artiques, 2014, *A Cautionary Tale: Spain's Solar PV Investment Bubble*, Winnipeg, Manitoba, Canada: International Institute for Sustainable Development, https://www.iisd.org/gsi/sites/default/files/rens_ct_spain.pdf. CC BY-NC-SA 4.0.

One aspect of FITs that received much positive attention was their potential for engaging community groups and small cooperatives that formed to invest in distributed energy resources, including wind, solar, and bioenergy. These projects resulted in combined benefits, including "boosting economic growth, strengthening market institutions, reducing poverty, creating new jobs, improving the local environment and health conditions, and mitigating global environmental risks."[53] Large numbers of such groups of small investors formed in part because the FIT compensation rates were generous, but

[53] Glemarec, Y., W. Rickerson, and O. Waissbein. 2012. *Transforming On-Grid Renewable Energy Markets: A Review of UNDP-GEF Support for Feed-In Tariffs and Related Price and Market-Access Instruments*. Washington, DC: Office of Scientific and Technical Information, Department of Energy. https://www.osti.gov/biblio/22090461. See also DeShazo, J.R., and R. Matulka. 2009. "Best Practices for Implementing a Feed-In Tariff Program." Los Angeles Business Council and UCLA Luskin Center for Innovation. https://innovation.luskin.ucla.edu/wp-content/uploads/2019/03/Best_Practices_for_Implementing_a_Feed-in_Tariff_Program.pdf.

another important factor was the publicly announced prices that gave people time to design, site, and develop projects.[54]

FINDINGS

This chapter has focused on trends affecting the context for the committee's examination of net metering: trends in the adoption of net metering as a billing mechanism to compensate BTM DG; the deployment of BTM DG in conjunction with net metering; and the cost of DG, most notably rooftop solar. The chapter also discusses tensions that have arisen as a result of these trends and states' responses in examining and adopting net metering variants. The committee makes the following findings:

Finding 3-1: By the end of the 1990s, net metering programs were offered to some or all electricity customers in 21 states. As of 2021, 40 states, the District of Columbia, and all 4 major U.S. territories had in place traditional net metering or successor tariffs.

Finding 3-2: Among various types of DG technologies, solar PV has been the dominant form of BTM technology adopted in conjunction with net metering, and almost all customer-sited PV installed to date has occurred where customers take service under net metering.

Finding 3-3: Net metering—or some form of compensation to a customer for its exports of electricity to the grid—has been broadly viewed as an enabler of the adoption of rooftop solar. Net metering is one factor, albeit sometimes a major factor, among a much larger set of considerations that have affected customers' interest in BTM technologies, including solar and storage.

Finding 3-4: Electricity prices, technology costs, public policies, DER deployment levels, industry structure, and other conditions in retail and wholesale electricity systems vary considerably across the jurisdictions where net metering has been adopted and have affected decisions in some places to transition net metering into variant approaches.

[54] Glemarec, Y., W. Rickerson, and O. Waissbein. 2012. *Transforming On-Grid Renewable Energy Markets: A Review of UNDP-GEF Support for Feed-In Tariffs and Related Price and Market-Access Instruments.* Washington, DC: Office of Scientific and Technical Information, Department of Energy. https://www.osti.gov/biblio/22090461, p. 13. See also DeShazo, J.R., and R. Matulka. 2009. "Best Practices for Implementing a Feed-In Tariff Program." Los Angeles Business Council and UCLA Luskin Center for Innovation. https://innovation.luskin.ucla.edu/wp-content/uploads/2019/03/Best_Practices_for_Implementing_a_Feed-in_Tariff_Program.pdf, pp. 3–5, 29–32.

Finding 3-5: Considerable uncertainty exists with respect to a number of issues relating to the future of net metering and with little standardization of policies across jurisdictions.

Finding 3-6: Policymakers in various jurisdictions are considering how and the extent to which existing and future wholesale and retail market conditions, including the costs and potential benefits associated with solar PV and other DER, should affect decisions about net metering, DG, and DER policies. It remains to be seen whether and how differences in market acceptance and growth in markets for DG will interact with and guide policies and regulatory changes.

Finding 3-7: Information about how BTM technologies that export power to the grid perform and interact with the grid is important for improving understanding of these interactions—both for local system operators but also for researchers, analysts, and others. Such information may be known to utilities, customers, and/or providers of BTM systems (e.g., the counterparty to a rooftop solar power purchase agreement) and is typically not available to others, often due to privacy, confidentiality, and/or technical reasons or even in a granular way to the customers with BTM technologies installed on their premises.

RECOMMENDATIONS

Recommendation 3-1: Utility regulators and decision makers of publicly owned utilities should work with stakeholders to explore and implement solutions to technical, legal, proprietary, or privacy concerns about the collection and release of information about the performance of behind-the-meter technologies and their interactions with the grid.

Recommendation 3-2: Data about participation in net metering and alternative and supplemental DG policies should be collected and reported so that policymakers can monitor how policies are interacting with energy market conditions in wholesale and retail, regulated and unregulated energy markets.

4

Economic Considerations Related to Net Metering

INTRODUCTION

The economics of net metering and its variants depend on the vantage point of different entities involved in the electricity system, which consist principally of:

- The distributed generation (DG) customer served on a net-metering tariff or other DG rates
- Other electricity customers (non-participants)
- The electricity system and utility that serves all those customers
- Society as a whole, including environmental and other effects that occur outside the boundaries of the electricity system

This chapter discusses the economic implications of net metering from these varied perspectives. These implications are based first and foremost on the benefits and costs of DG. From the perspective of the DG customer, the economic implications depend on how the level of compensation through net metering, along with other benefits, compares to the costs of DG ownership. From the perspective of the utility and non-participating customers, the economic implications depend on how the DG customer's bill savings under the net metering tariff compare to the costs that the utility avoids by virtue of the DG systems. From the perspective of society as a whole, the economic implications further depend on environmental costs and other externalities of electricity production, which may be reduced by non-emitting DG resources. In addition to aggregate benefits and costs of net metering, there are distributional effects in terms of who, specifically, receives the benefits and who bears the costs. This chapter addresses those distributional effects broadly between DG customers and non-DG customers overall, while the next chapter, about equity considerations, addresses distributional effects among the distinct socio-economic groups that make up the utility's customer base.

When evaluating these economic impacts, one must be careful to distinguish between the impacts of net metering and the impacts of DG. Net metering is a billing mechanism for customers with DG. Its function is to establish a customer's billing determinants (e.g., the quantity of kilowatt-hours) and the rate at which they are billed each month. As such, net metering is intended to provide a direct economic value to participating customers by compensating them for their DG production, based on the underlying rate structure. It also has distributional effects in terms of how the utility's revenues are recovered across its customer base. However, net metering has no intrinsic benefits or costs from a societal perspective.[1] Rather, those societal benefits and costs are properties of the DG adoption induced by net metering.

Insofar as net metering incentivizes customers to adopt clean DG, it can be evaluated as a clean energy policy support mechanism, and measured against some counterfactual—that is, what would have happened in the absence of net metering customers adopting those DG resources? Within this framework, there may be costs to net metering insofar as it induces either more or less DG than would be optimal, from a societal or power-system perspective. The economic merits of net metering can also be compared to other forms of clean energy policy support, which might deliver the same level of benefit at a lower or higher public cost or with more or less equitable distribution of those costs.

Specific topics examined in this chapter include:

- The economic implications of net metering and its variants from the perspective of DG customers
- The societal perspective, in which net metering and its variants can be evaluated in terms of economic efficiency
- The non-participant and utility perspective, where the effects of net metering and its variants on non-participant impacts are evaluated principally in terms of changes in retail rates
- A discussion of the many benefits and costs of DG that are relevant to assessing the economic efficiency and rate impacts of net metering and its variants
- Findings and recommendations

DG CUSTOMER PERSPECTIVE

Customer adoption of DG technologies is a complex behavior, driven by a combination of economic and non-economic factors. Utility bill savings are one key consideration in that adoption decision, though other factors may also play a central role, including the ability to hedge against future increases in electricity prices, incentives (e.g., the federal investment tax credit and renewable energy certificates), customer perceptions of the

[1] One minor exception would be incremental metering and billing costs associated with net metering.

utility, environmental views, status value, social influences,[2] and potential reliability and resilience benefits for DG resources that can provide backup power during grid interruptions, among other factors.[3] For example, customers who identify as environmentalists may be willing to invest in solar in part because they value the opportunity to behave consistently with their values and protect the environment. Customers may also be willing to pay for solar given its ability to signal certain characteristics about themselves (e.g., environmentalism), a phenomenon known as conspicuous conservation.[4] The impact of DG compensation rates on adoption behavior thus depends in part on the strength and relevance of those myriad other factors. Other attributes of the DG compensation structure, such as its simplicity or stability, may also affect its influence on customer adoption behavior.

Compensation for behind-the-meter (BTM) DG production typically includes two components: electricity bill savings from reducing the quantity of electricity consumed from the grid (area B in Figure 4-1) plus any bill credits provided for surplus DG production exported to the grid (area A), which are applied against the utility billing charges for the customer's residual usage (Area C), and may be carried over to the following billing period, in whole or in part, depending on the provisions of the net metering or DG compensation tariff. A residential solar DG system sized to generate an amount equal to the host customer's annual electricity consumption, and without battery storage, will typically export 60–70 percent of generation over the course of the year.[5] Thus, the rules and pricing applied to exported generation can have a significant impact on the

[2] Bollinger and Gillingham (2012) find that for each additional residential rooftop installation in a California zip code, the likelihood of another installation increases by 0.78 percent. Similar effects have been observed in the United Kingdom (Richter 2014). This effect tends to be larger with greater public visibility and lower physical proximity (Wolske et al. 2020). Bollinger, B., and K. Gillingham. 2012. "Peer Effects in the Diffusion of Solar Photovoltaic Panels." *Marketing Science* 31(6):900–912; Richter, L. 2014. "Social Effects in the Diffusion of Solar Photovoltaic Technology in the UK." EPRG Working Paper 1332. University of Cambridge. https://www.eprg.group.cam.ac.uk/wp-content/uploads/2013/12/1332-Final.pdf; Wolske, K.S., K.T. Gillingham, and P.W. Schultz. 2020. "Peer Influence on Household Energy Behaviours." *Nature Energy* 5:202–212.

[3] Moezzi, M., A. Ingle, L. Lutzenhiser, and B. Sigrin. 2017. *A Non-Modeling Exploration of Residential Solar Photovoltaic (PV) Adoption and Non-Adoption*. Golden, CO: National Renewable Energy Laboratory. https://www.nrel.gov/docs/fy17osti/67727.pdf.

[4] Griskevicius, V., J.M. Tybur, and B. Van den Bergh. 2010. "Going Green to Be Seen: Status, Reputation, and Conspicuous Conservation." *Journal of Personality and Social Psychology* 98:392–404. The same phenomenon has been observed with hybrid car adoption: Prius buyers are willing to pay $430–$4,200 more than equivalent hybrid models that lack the distinctive body styling of the Prius for status signaling; see Sexton, S.E., and A.L. Sexton. 2014. "Conspicuous Conservation: The Prius Halo and Willingness to Pay for Environmental Bona Fides." *Journal of Environmental Economics and Management* 67:303–317.

[5] Forrester, S., G. Barbose, and C. Miller. 2022. "Private vs. Public Value of U.S. Residential Battery Storage Operated for Solar Self-Consumption." *iScience* 25(8):104714.

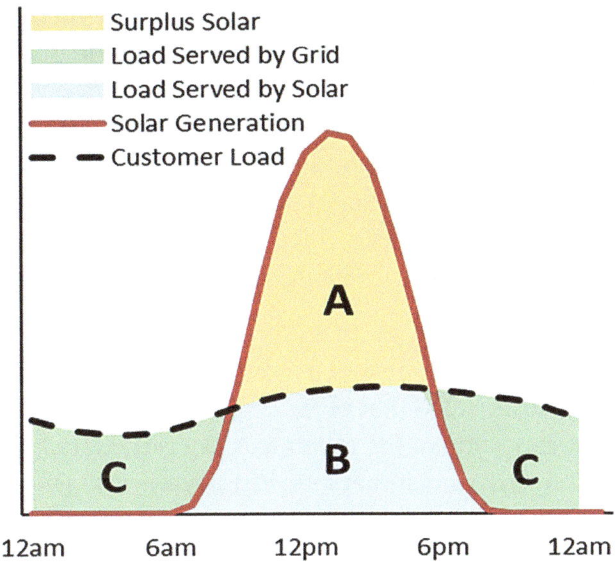

FIGURE 4-1 Illustration of billing determinants for customers with behind-the-meter solar DG.
SOURCE: S. Forrester, G. Barbose, and C. Miller, 2022, "Private vs. Public Value of U.S. Residential Battery Storage Operated for Solar Self-Consumption," *iScience* 25(8):104714, https://doi.org/10.1016/j.isci.2022.104714. BY-NC-ND 4.0.

customer-economics of investing in solar DG and by extension, on the level of adoption induced and the resulting benefits and costs for the broader electric systems.

Traditional net metering credits all DG production (areas A and B in Figure 4-1) based on the volumetric energy rate of the customer's retail electricity tariff. The compensation provided through net metering is thus inherently linked to the structure and pricing of the customer's underlying retail rates. Electricity tariffs that charge customers primarily through volumetric energy-based charges, with little or no fixed customer charges or demand charges, provide higher levels of compensation for DG production, all else being equal.[6] Electricity tariffs with time-varying (e.g., time-of-use or hourly) energy rates may provide higher or lower levels of DG compensation, depending on how well the timing of DG production coincides with peak prices.

Customers considering DG have three broad options: (1) continue as a full-service customer of the electric utility without DG; (2) become a partial-service customer,

[6] Electricity tariffs that charge customers in volumetric rates for their use of the distribution and transmission system may also enable DG customers to avoid paying for their use of or share of support for this infrastructure, which provides broad "resilience" service for all customers as long as they are connected to the grid.

self-generating a portion of their own needs and relying on the utility as necessary for the balance of their needs; and (3) fully disconnect from the grid, self-supplying the entirety of their electricity needs with DG and potentially battery storage or fossil back-up generation, or both. In the case of BTM solar generation, the customer-economics of these three options has evolved over time, as a result of changes in both technology costs as well as net metering rules and rate structures. This evolution and possible futures are illustrated by the cases shown in Figure 4-2, which, it should be noted, present points along a continuum and not a menu of discrete choices.

- **Case 1:** Under traditional net metering structures, the customer-economics of becoming a partial requirements customer with solar DG depends on how the levelized cost of solar DG (net of tax credits and other incentives) compares to the volumetric portion of the customer's electricity rate, given that all DG production is compensated at that rate. Historically, solar DG was more expensive than volumetric rates in most regions, and so the least-cost option for most customers was to remain as full-service utility customers.
- **Case 2a:** As solar DG costs have declined, partial requirements service under net-metering tariffs has increasingly become economic, particularly in regions with high solar insolation, high electricity rates that consist primarily of volumetric charges or other attractive state- or utility-level incentives. Those states with the highest penetration levels (as shown previously in Figure 3-6) are indicative of where the customer economics for solar DG has been most favorable.
- **Case 2b:** Utilities and regulators in several states have considered, and in some cases implemented, modifications to the compensation structure for solar DG. Two modifications that have been frequently considered are net billing structures and time-varying rates. Both of these methods are intended to incentivize customers to shift the timing of their solar DG production—in the case of net billing rates, to those times when it can be used to directly serve onsite load, and in the case of time-varying rates, to those times when it has the greatest value to the electricity system. Under these types of compensation structures, partial requirement service may be most economical for customers that install battery storage along with solar DG, to store solar production to offset consumption at those times of day that maximize the customer's utility bill savings.[7]
- **Case 3:** Utilities and regulators in some jurisdictions have also considered modifications to DG compensation structures consisting of either higher fixed charges

[7] Under net billing, customers will use batteries to maximize self-supply by storing excess solar production when it exceeds consumption, provided that the compensation for export is lower than the retail price for electricity. Under time-varying rates, customers will charge their batteries from solar DG (or potentially from the grid, if allowed) when retail prices are low and discharge their batteries when retail prices are high.

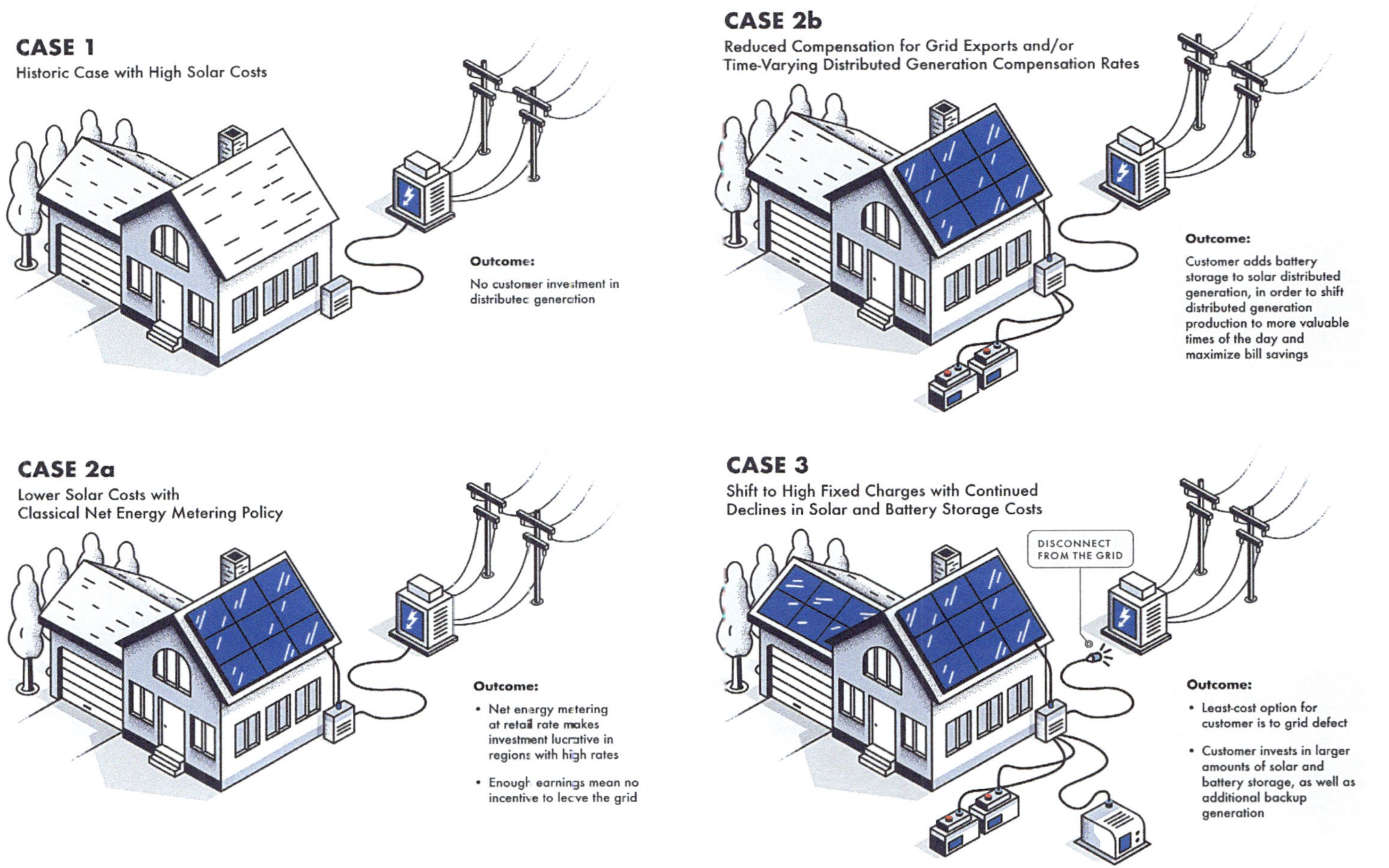

FIGURE 4-2 Economic cases of net metering from the customers perspective.

or grid-access fees for solar DG customers that they cannot avoid. As a result, where those billing components are high enough, some customers may find it most economical to disconnect from the grid and provide their electricity service needs solely through some combination of solar DG, batteries (or electric vehicles with bidirectional charging), and backup generators. In disconnecting from the grid, the customer's resources would also be unavailable to provide services to the grid. Box 4-1 discusses the customer decision-making process to disconnect from the grid in more detail.

THE SOCIETAL PERSPECTIVE: ECONOMICALLY EFFICIENT PRICING

As discussed in Chapter 7, utilities and regulators balance various competing objectives when developing retail electricity rates. One of those key objectives is to send price signals to customers so that they can make economically efficient decisions about their consumption. According to standard economics theory, economic efficiency dictates that prices should be set equal to social marginal cost:

- *Social Marginal Cost (SMC)* is the incremental cost to the company supplying an additional unit of the good or service *plus the costs of any externalities*, such as pollution, that result from that incremental supply. The concept of externality costs is described further later in this chapter.

If price reflects the SMC, then consumers deciding whether to buy the product are accurately comparing the value they would receive from the product with society's cost of providing an additional unit of it. In contrast, prices below the SMC will lead consumers to purchase some quantity of the good that they value at less than society's cost of supplying them, whereas prices above SMC will lead consumers to forgo consumption when the value they would receive is greater than society's cost of supply. Both cases result in reduced economic value.[8]

This concept of SMC can be differentiated from the private marginal cost incurred by the supplier of the good or service:

- *Private Marginal Cost (PMC)* is the incremental cost to the company supplying an additional unit of the good or service, *excluding costs of any externalities.*

[8] In economic textbook terms, this is referred to as "deadweight loss."

BOX 4-1
UNDERSTANDING CUSTOMER DECISIONS TO DISCONNECT

The dynamics of a customer's decision to deploy BTM DG can also apply to a decision to disconnect from the electricity grid. A combination of technologies in off-grid systems including photovoltaic (PV), batteries, and generators or cogeneration now make it technically possible for customers to entirely disconnect from the grid and cease being electric utility customers by providing all of their own electricity (referred to by some as "grid defection"). For example, in northern Michigan, both residential (Kantamneni et al. 2016) and commercial (Peffley and Pearce 2020) customers can produce their own electricity less expensively than the grid offers, and therefore have an incentive to disconnect from the grid.

While the circumstances that may drive customers to disconnect, and the instances of actual disconnection are currently few, understanding this dynamic is important for policymakers and regulators so that they design net metering variants that discourage uneconomic disconnection and leverage the BTM DG for the benefit of the electricity system (i.e., where customers that disconnect from the grid pay more for their electricity needs than it would cost the utility to serve them). A customer has the incentive to disconnect from the grid when they can receive comparable or better service from a DG system (PV, storage, and perhaps fossil backup) at less than the total cost of their pre-DG utility bill over the useful life of the DG system. Net metering compensation levels can be a factor in a customer's decision to disconnect. For example, traditional net metering of BTM DG provides a much more convenient and less capital-intensive method for customers to reduce their electricity bills using rooftop solar and can avoid uneconomic outcomes (for all customers) when retail rates are aligned with costs to serve. To the extent that utility cost recovery occurs more through fixed charges, variable charges will decrease along with the resulting net metering compensation for the customer.

Grid disconnection may be more attractive for customers who are willing to compromise reliability for lower costs (e.g., having less battery storage and no fossil backup) or are in isolated areas where they would be required to pay additional costs to be connected to the grid. In cases of economic grid disconnection, where the customer is able and willing to self-supply at a cost less than the utility would bear to provide service, then grid disconnection or defection may be an acceptable outcome.

References

Kantamneni, A., R. Winkler, L. Gauchia, and J.M. Pearce. 2016. "Emerging Economic Viability of Grid Defection in a Northern Climate Using Solar Hybrid Systems." *Energy Policy* 95:378–389.

Peffley, T.B., and J.M. Pearce. 2020. "The Potential for Grid Defection of Small and Medium Sized Enterprises Using Solar Photovoltaic, Battery and Generator Hybrid Systems." *Renewable Energy* 148:193–204.

In general, SMC is greater than PMC. Environmental damage costs can be internalized through regulatory mechanisms, such as emission control requirements or cap-and-trade programs, at least partially closing the gap between SMC and PMC. However, to the extent externalities remain, pricing at the PMC will tend to induce an economically inefficient over-consumption of the good or service from a societal perspective.

Within the context of electricity pricing, economic efficiency dictates that volumetric energy prices faced by consumers should reflect the SMC, that is, the incremental costs of producing and delivering electricity to the point of use, including environmental externalities.[9] These marginal costs differ between the short and long-run, as certain types of costs that are fixed in the short run may be variable over longer timeframes. In practice, electricity prices usually do not accurately reflect this SMC. This can occur for several reasons, which tend to work in opposite directions. First, under cost-of-service ratemaking, electricity prices are set to recover only the costs incurred directly by the utility in producing and delivering electricity, and therefore exclude any externalities. At the same time, electricity rates, particularly for residential customers, often recover both fixed and variable costs through volumetric energy charges.[10] Those prices may therefore be greater than the utility's PMC, and depending on the size of the externalities, potentially greater than the SMC as well. The combination of these effects differs across the United States (e.g., environmental externalities in regions that rely heavily on coal-fired electricity generation are different than in regions that rely predominantly on low- or zero-carbon electricity). As a result, the marginal electricity prices faced by residential customers may either be higher or lower than the SMC of producing and delivering electricity.

The implication of the above discussion for net metering, which compensates customers for DG production based on the volumetric energy price, is that it may either overcompensate or under-compensate DG, depending on whether prices are above or below SMC. That misalignment, in turn, can lead to either an over- or under-investment of DG resources on the part of customers, from a societal perspective.

[9] To be sure, there are a variety of limitations to and constraints on regulators' ability to price electricity at SMC. First, externality costs must be estimated, and those estimates are often uncertain and may evolve over time, for reasons discussed later in this chapter, in the section titled "Environmental and Other Externalities." Second, regulators have a variety of objectives in setting electricity rates, including simplicity and predictability, which must be balanced against the goal of economic efficiency. Those tradeoffs are discussed further in Chapter 7. Third, pricing electricity at the SMC will typically result in either an under- or over-recovery of utility costs, depending on how SMC compares to utility's average cost of service. Other revenue recovery or crediting mechanisms are therefore needed to ensure that the utility is able to recover its costs.

[10] This has been done traditionally for many policy reasons including, but not limited to, simplicity of rate design, encouraging conservation, and sometimes in an effort to avoid inequitable outcomes such as imposing high fixed charges on lower-income customers.

In regulatory settings involving net metering and compensation levels for DG, the concept of SMC is sometimes discussed instead in terms of "avoided costs," that is, the net costs avoided by each unit of DG production. These cost impacts are discussed later in this chapter, along with key considerations in how they can be quantified for the purpose of evaluating and designing DG compensation rates. The key question, from the perspective of the economic efficiency of net metering, is whether the volumetric energy component of the customer's retail electricity tariff is greater or less than the total of those avoided costs, including any externalities.

As described in Chapter 1, a variety of options exist for utilities and regulators to better align DG compensation with SMC/avoided costs, in cases where net metering either over- or under-compensates DG production. Some of those options involve changes to the underlying retail electricity rate structure, which affects compensation for both the portion of DG production exported to the grid as well as the portion self-consumed by the host customer. In cases where volumetric energy prices exceed the SMC (including externalities), higher customer charges or demand charges could help to bring volumetric energy rates into closer alignment with SMC. And in all cases, even where volumetric energy prices closely approximate SMC on an annual average basis, introducing time-varying electricity prices that closely mirror the temporal profile of SMCs will provide more economically efficient compensation for DG of all types and will likely encourage DG production during the times of day when it is most valuable. Time-varying rates have taken on even greater relevance as battery storage has become less expensive, by providing customers with more flexibility in the timing of their consumption and production.

As discussed in Chapter 7, utility ratemaking is guided by a number of principles, which may at times conflict with the goal of setting economically efficient rates. Thus, while utilities and regulators may strive to implement prices that better correspond to SMC, there are often limitations to doing so.[11] The two primary variants of net metering—net billing and buy-all or sell-all rates[12]—offer a means to implement more efficient DG compensation rates without relying solely on changes to underlying retail rate structures. Both variants allow utilities and regulators to specify the price paid for DG production, which can be set directly at the SMC, inclusive of externalities and with the appropriate temporal and spatial differentiation, to capture key variations in marginal costs.

[11] For example, volumetric energy prices based on SMC may lead to an over- or under-recovery of the utility's cost of service, depending on the level of fixed customer charges imposed.

[12] See Figure 1-1.

In the case of buy-all and sell-all rates, which apply to the totality of DG production, setting the tariff for both consumption (buy-all) and production (sell-all) at a time- and locationally varying SMC provides an economically efficient price signal for DG investment and operation. Net billing sets only the rate for exported DG production, and thus the effective compensation rate that a customer considers when making the DG investment decision will be some average of the net billing rate for exports and the volumetric energy rate of the customer's retail rate. Setting the net billing rate at the SMC will thus bring the effective DG compensation rate closer to, but not equal to, the economically efficient level. The asymmetric pricing structure of net billing, where one price applies to the portion of DG self-consumed by the customer and another price applies to the portion exported to the grid, may incentivize customers to install battery storage for the purpose of capturing the value differential between self-consumption and export prices.[13]

UTILITY AND NON-PARTICIPANT RATE IMPACTS

Separate from any assessment of its economic efficiency, net metering and other DG compensation mechanisms also have distributional impacts in terms of who receives the associated benefits and who bears the costs. Among the different entities affected, the electric utility is impacted primarily in two ways: (1) first, by the effects of DG resources on the utility's costs (both positive and negative) and the attendant impacts on its revenue requirements; and (2) second, by any lost revenues associated with retail electricity sales displaced by BTM DG resources. These impacts, in turn, flow to utility shareholders and utility ratepayers in distinct ways. Shareholder impacts are discussed briefly in Chapter 7, while the remainder of this section focuses on the impacts to utility ratepayers—in particular, the set of utility ratepayers not enrolled in net metering or other DG tariffs, often termed "non-participants."

Non-participant impacts are typically defined in terms of the effect on utility rates: that is, whether and to what extent the DG compensation mechanism leads to an increase or a decrease in electricity rates for non-participants. As a general matter, the rate impacts of any DG compensation mechanism are a function of how the reduction in the utility's revenue requirements compares to lost revenues. If lost revenues are greater than the reduction in revenue requirements, the utility may need to raise its prices or institute other

[13] Some assert that this form of "regulatory arbitrage" (Green, R., and I. Staffell. 2017. "'Prosumage' and the British Electricity Market." *Economics of Energy & Environmental Policy* 6(1):33–50) is unlikely to represent an economically efficient application of battery storage (Forrester, S., G. Barbose, and C. Miller. 2022. "Private vs. Public Value of U.S. Residential Battery Storage Operated for Solar Self-Consumption." *iScience* 25(8):104714).

charges in order to recover its costs; rates and bills for non-DG customers will therefore rise. Conversely, if the reduction in revenue requirements exceeds lost revenues, then rates and bills for non-participants may decline.

In assessing non-participant rate impacts of the DG compensation mechanism, a key definitional issue is the time horizon of the analysis. In the short run, a utility may have large, fixed costs associated with depreciation and interest expenses of past investments. During this period, DG production may have a relatively small effect on utility revenue requirements, because they are largely based on past investments. As a result, the lost revenue effect outweighs the cost reduction effect.

Over the longer term, BTM DG may defer or avoid future capital expenditures (e.g., related to new generation capacity or transmission and distribution equipment), or in some cases, DG may impose additional costs (e.g., distribution network upgrades needed to integrate distributed resources). However, such effects are highly specific to each situation, as discussed further in the next section. Reductions or deferrals in future capital expenditures will further reduce utility revenue requirements; all else being equal, the non-participant rate impacts will be correspondingly lower and, in some circumstances, could be negative (i.e., non-participants see rate decreases).[14]

Under net metering, assessing the non-participant rate impacts can be viewed as a comparison between the volumetric energy rate (i.e., the price at which DG production is compensated under net metering) and the utility's (private) marginal costs during the hours when DG production occurs. The non-participant rate impacts of net metering are thus highly dependent on the customer's underlying retail rate structure as well as the temporal profile of the utility's marginal costs. Under the predominant rate structure for residential customers consisting of flat (non-time-varying) volumetric energy charges, with small fixed customer charges and no demand charge, the volumetric energy rate will generally exceed the utility's marginal costs in most hours.[15] Net metering tariffs coupled with this type of rate structure will therefore often lead to some rate increase for

[14] Satchwell, A., A. Mills, and G. Barbose. 2015a. "Quantifying the Financial Impacts of Net-Metered PV on Utilities and Ratepayers." *Energy Policy* 80(May):133–144.

[15] The large, fixed costs of electricity grids are associated with the characteristics of its historically being a natural monopoly. A natural monopoly is a market in which total costs of supplying the commodity are lower if it is supplied completely by one firm, rather than by multiple firms. Typically, a natural monopoly occurs when the costs of new entry are high because of the need for a large, fixed investment, with low incremental costs and limited substitutable goods. As a result of this cost structure, the marginal cost of supplying additional output is below or very low compared to the average cost, which averages in all the fixed costs.

non-participants, though any such analysis requires consideration of the specific timing of DG production and utility marginal costs in those hours.[16]

Non-participant rate impacts can be mitigated either through changes to the underlying rate structure or by transitioning away from traditional net metering. In general, rate structures with time-varying volumetric energy prices closely coinciding with the temporal profile of the utility's marginal costs will tend to minimize non-participant rate impacts.[17] Setting volumetric energy rates at the utility's marginal cost generally necessitates the use of fixed customer charges or demand charges in order to ensure utility cost recovery. However, increasing fixed charges can also mute the price signals that encourage customers to pursue energy efficiency and exacerbate equity issues for low-income customers by raising their effective rate (e.g., the less electricity a customer uses the more she pays per kWh).[18] Alternatively, non-participant rate impacts also can be mitigated by replacing traditional net metering with either net billing or buy-all and sell-all rates, setting the price for DG production to reflect the utilities' marginal costs, or by applying additional fees or charges on net-metering customers. As noted previously, in the case of net billing tariffs, customers may install battery storage so that they can use their DG production to meet the onsite loads. In doing so, most DG production will continue to be compensated at the volumetric rate of the customer's retail electricity tariff, leading to similar non-participant rate impacts as would occur under traditional net metering.

Important to note is that the goals of minimizing non-participant rate impacts and maximizing economic efficiency stand in some degree of conflict with one another. This is because non-participant rate impacts are minimized when DG compensation is based on PMC, while economic efficiency is maximized when DG compensation is based on SMC, which exceeds PMC. Setting DG compensation at the socially optimal level, which reflects externalities, will thus often engender some rate increase for non-participants. This represents one potential point of tension that regulators and utilities must consider in setting electricity rates. (See Chapter 7 for further discussion.)

[16] Lawson, A.J. 2019. *Net Metering: In Brief.* CRS Report R46010. Washington, DC: Congressional Research Service.

[17] Overall, the impact of time-varying rates on customers' bills will depend on their ability to shift the timing of their consumption from higher to lower cost time periods. As discussed in Chapter 5, certain groups of customers, such as low-income customers, may have a limited ability to respond to time-varying rates.

[18] Pearce, J.M., and P.J. Harris. 2007. "Reducing Greenhouse Gas Emissions by Inducing Energy Conservation and Distributed Generation from Elimination of Electric Utility Customer Charges." *Energy Policy* 35:6514–6525.

Finally, an essential consideration when evaluating the non-participant rate impacts of net metering or other DG compensation mechanisms is the absolute magnitude of the impact, which depends critically on the level of DG penetration in any given region. As shown in Chapter 3, net-metered solar penetration at the end of 2021 in most states was below 1 percent of all residential customers, and in all but 4 states was below 5 percent. In regions with low levels of net metering participation, effects on non-participants' bills will be small, and in some cases de minimis, regardless of how widely DG compensation rates diverge from the utility's marginal costs. As net metering participation rates climb, however, rate impacts may become more salient.

Figure 4-3 presents a simple illustrative schematic showing how rate impacts depend on the interplay between distributed solar penetration levels (denominated as total distributed solar generation as a percent of total electricity consumption), the DG

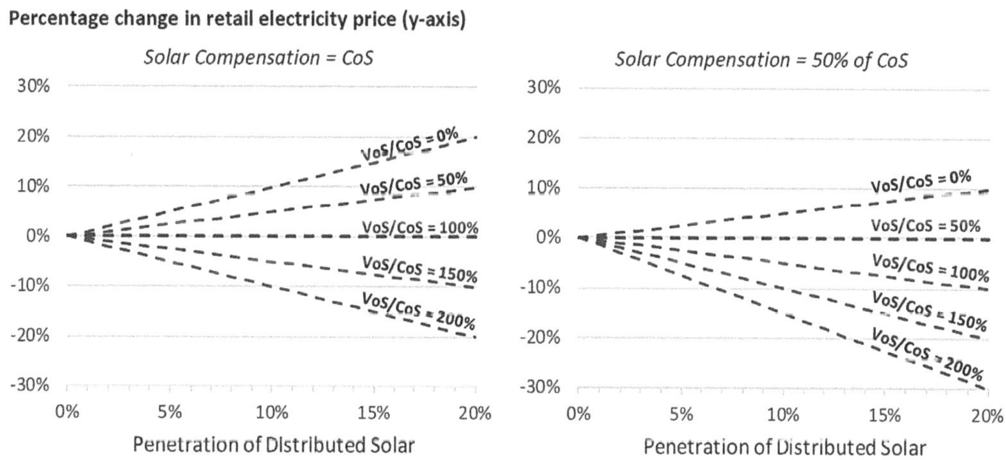

FIGURE 4-3 Impacts of distributed solar average retail electricity prices: a simple model of underlying drivers.

NOTES: CoS refers to the utility's average cost of service, VoS refers to the avoided utility costs from solar, and Penetration refers to distributed generation as a percentage of total customer load. Solar Compensation is denoted as a fraction of the CoS and represents the average compensation per kWh of solar generation provided to the customer, net of any fees and inclusive of bill savings from directly offsetting load. The dashed lines are meant to capture a wide range in avoided costs (characterized as VoS), expressed as a percentage of CoS, ranging from a VoS of zero to a VoS that is double the utility's CoS. Most avoided cost/VoS studies fall somewhere in that range (ICF 2018).[a]

[a] ICF Consulting. 2018. Review of Recent Cost-Benefit Studies Related to Net Metering and Distributed Solar. Washington, DC: ICF Consulting. https://www.energy.gov/sites/default/files/2020/06/f75/ICF%20NEM%20Meta%20Analysis_Formatted%20FINAL_Revised%208-27-18.pdf.

SOURCE: G. Barbose, 2017, *Putting the Potential Rate Impacts of Distributed Solar into Context*, Berkeley, CA: Lawrence Berkeley National Laboratory, https://emp.lbl.gov/publications/putting-potential-rate-impacts.

compensation rate (denominated in normalized terms as a percentage of the utility's average cost of service, or CoS), and the utility's avoided costs, denoted in Figure 4-3 as the value of solar (VoS) to the utility (and equivalent to the concept of private marginal cost introduced earlier). For example, compensating DG at a price equal to the utility's full average CoS (as under a traditional net metering tariff with flat volumetric rates) and where the utility avoided cost (or VoS) is just 50 percent of the CoS (as might occur if half of the utility's costs are fixed and half are variable) would result in about a 10 percent increase in average retail rates once penetration levels reach 20 percent of customer load.

ESTIMATING THE ELECTRICITY SYSTEM AND BROADER SOCIETAL IMPACTS OF DG

Estimating the benefits and costs of DG first requires identifying a comprehensive list of impacts, then evaluating the incremental value of each of these impacts relative to a counterfactual scenario without a particular DG resource. For example, the carbon emissions reductions from installing additional DG are equal to the amount of generation from carbon-emitting resources that DG displaces. Calculating this requires an estimate of how much carbon-emitting generation would have occurred absent the DG. This section first explains the key considerations for a robust analysis, identifies the major categories of DG impacts, and explains how each of these major impacts of DG could be calculated.

Key Considerations in Evaluating Future DG Impacts

A robust counterfactual-based DG analysis is location-specific, contains adequate temporal detail, applies best available load and power supply forecasts to estimate future impacts of DG, accounts for relevant policy goals, and helps determine long-term benefits and costs of DG through its lifetime. These considerations are explained in detail next.

Regional specificity: DG installation in a specific location will primarily affect electricity purchase, resource build, and grid operation of the utility that serves that location.[19] The baseline should therefore be constrained to the utility territory in which the DG is going to be installed. Moreover, each utility has its own unique rate structure which is the foun-

[19] DOE. 2007. *The Potential Benefits of Distributed Generation and the Rate-Related Issues That May Impede Its Expansion: Report Pursuant to Section 1817 of the Energy Policy Act of 2005.* Washington, DC: Department of Energy. https://www.energy.gov/sites/prod/files/oeprod/DocumentsandMedia/1817_Report_-final.pdf.

dation for the net metering tariff it offers. Multiple utility territories (e.g., in organized markets with central dispatch of generation) could be combined to develop a common counterfactual with careful consideration of the load each utility serves, its electricity grid needs, clean energy legislation and regulation, and other policy goals for which each utility is accountable. There may be instances where DG affects specific parts of the distribution grid (e.g., individual feeders and circuits) being analyzed, and in that case the geographical boundary of the baseline will be smaller than a utility service territory.

Temporal detail: Energy use varies with time, and the effect of a DG resource is, among other things, a function of a region's energy use profile and the DG's generation loadshape, which vary by hours of the day and seasons of the year. Moreover, in increasingly renewable grids, energy prices and carbon emissions vary significantly with time (related to what type of generation is being displaced).[20] This means that the economic and environmental benefits of DG will also vary with time. The counterfactual baseline scenario may therefore include this level of temporal detail to be able to capture the different streams of DG benefits. Many states, such as California[21] and New York,[22] account for this time-varying impact of different DG and DER.

Load forecast: An estimate of current and future demand for electricity consumption is needed to determine what future loads DG will meet and how these loads would be met absent the DG. The forecast will need to consider any shifts in seasonal peaks, as when a utility foresees moving from winter to summer peaking or vice versa.

Fuel and technology cost forecasts: Fuel and technology forecasts are necessary for rigorous estimates of avoided energy generation, capacity, and environmental costs, each of which is explained later in this section. Fuel forecasts—generally natural gas and coal—are necessary to estimate the costs of future energy generation from fossil resources, as the costs of operating these fossil resources are a key determinant of wholesale prices of

---

[20] Kaffine, D.T., J.M. Brannin, and J. Lieskovsky. "Emissions Savings from Wind Power Generation in Texas." *The Energy Journal* 34(1):155–175.

[21] See CPUC hourly avoided costs calculator: CPUC. 2021b. "IDER Cost-Effectiveness, Avoided Cost Calculator." https://www.cpuc.ca.gov/industries-and-topics/electrical-energy/demand-side-management/energy-efficiency/idsm.

[22] See New York's solar value stack calculator: NYSERDA (New York State Energy Research and Development Authority). 2022b. "Solar Value Stack Calculator." https://www.nyserda.ny.gov/All-Programs/NY-Sun/Contractors/Value-of-Distributed-Energy-Resources/Solar-Value-Stack-Calculator.

electricity. Fossil fuel cost forecasts also can include future carbon emissions liabilities.[23] Similarly, the forecasts of clean energy technology (solar, wind, and battery storage) installed costs are necessary for determining the future costs of clean energy generation and capacity. Established sources of these forecasts include, but are not limited to, National Renewable Energy Laboratory's Annual Technology Baseline (ATB)[24] and the Energy Information Administration's short term energy outlook.[25]

Applicable policy goals: The baseline economic conditions related to how a utility plans to meet its future electricity demand depend in part on the policies, especially clean energy mandates, with which it needs to comply. Currently, 29 states plus the District of Columbia have some form of renewable portfolio standards (RPS) or clean energy standards (CES) that requires utilities to meet a predetermined fraction of electricity sales from renewable or zero carbon resources, or both, in some cases with a pre-determined set-aside for DG.[26] Counterfactuals of resource-build and operation will differ for those states that have such policies versus those that do not. Therefore, while there may be more non- or low-emitting generation in states with strong clean energy goals, the carbon emission reduction of incremental DG in those states may be less than in states and grids without such goals.

Time horizon of analysis and the quantity of DG being analyzed: The amount of DG being analyzed and the time horizon over which it is analyzed dictate whether its impacts are only on the operating margin or whether it significantly affects the make-up of the grid, in terms of both generation resources as well as transmission and distribution infrastructure (discussed in Chapter 6). For example, a large amount of distributed solar and storage adoption may mean that less supply-side solar and storage will be contracted. The time horizon of the counterfactual should be equal to or longer than the estimated useful life of the DG being evaluated.

---

[23] Allen, M. 2003. "Liability for Climate Change." *Nature* 421(6926):891–892; Gundlach, J. 2020. "Climate Risks Are Becoming Legal Liabilities for the Energy Sector." *Nature Energy* 5:94–97; Heidari, N., and J.M. Pearce. 2016. "A Review of Greenhouse Gas Emission Liabilities as the Value of Renewable Energy for Mitigating Lawsuits for Climate Change Related Damages." *Renewable & Sustainable Energy Reviews* 55:899–908; Krane, J. 2017. "Climate Change and Fossil Fuel: An Examination of Risks for the Energy Industry and Producer States." *MRS Energy & Sustainability* 4:E2.

[24] NREL. 2022a. "Annual Technology Baseline (ATB): 2022 Electricity ATB Technologies." https://atb.nrel.gov/electricity/2022/technologies.

[25] EIA. 2022f. "Short-Term Energy Outlook." https://www.eia.gov/outlooks/steo/report/natgas.php. For additional information on CONE see the 2021 National Energy Technology Laboratory (NETL) report: NETL. 2021. "Energy Technology Systems and Market Analyses." https://netl.doe.gov/sites/default/files/netl-file/21TPG_Brewer.pdf.

[26] North Carolina Clean Energy Technology Center. 2022a. "Database of State Incentives for Renewables & Efficiency® (DSIRE)." https://www.dsireusa.org.

Discount rates: Discount rates reflect the time value of money and are used when valuing the benefits and costs of BTM DG over time. Discount rates have the effect of placing greater value on earnings or savings incurred sooner rather than later. The higher the discount rate, the lower the value placed on future impacts, compared to present-day impacts.

The appropriate discount rate depends on the perspective of the entity being analyzed. The discount rate for a utility is often set at the utility's weighted average cost of capital (WACC), or its "time value of money." If a utility made investments at a future date, the utility would forego some return on investment that it otherwise could have realized if it made the investment now. The WACC for U.S. electric utility companies has generally been around 8 percent.[27] The level depends in part on interest rates. Societal costs are generally evaluated with a lower social discount rate, indicating that future societal damages due to present-day carbon and air pollution emissions are not discounted as much as private investment. Social discount rates are usually in the range of 1 to 3 percent.

Ideally, analysts could apply the relevant discount rate to each stream of avoided costs or benefits. When calculating the present value of a utility's future avoided wholesale electricity purchases, a utility WACC can be used. When calculating the present value of the social cost of carbon, a societal discount rate should be applied. The choice of discount rate has equity implications regarding the value given to impacts on future generations, as discussed in Chapter 5.

MAJOR ELECTRIC SYSTEM AND SOCIETAL IMPACTS OF DG

Many studies exist that explain each stream of DG benefits and costs in detail, along with available methods to quantify them.[28] This section summarizes existing research, focuses on the major streams of benefits and costs of DG, defines them, presents the general methods for quantification, and recommends key considerations when estimating these

[27] The U.S. average ROR from 53 jurisdictions, which are all around 8 percent, is available at Quackenbush, J.D. 2019. *Cost of Capital and Capital Markets Primer for Utility Regulators*. Washington, DC: National Association of Regulatory Utility Commissioners, p. 25. https://pubs.naruc.org/pub.cfm?id=CAD801A0-155D-0A36-316A-B9E8C935EE4D.

[28] CPUC. 2022c. "Decision on Transportation Electrification Policy and Investment." Proposed Decision, Rulemaking 18-12-006. https://docs.cpuc.ca.gov/PublishedDocs/Efile/G000/M497/K622/497622010.PDF; Hayibo, K.S., and J.M. Pearce. 2021. "A Review of the Value of Solar Methodology with a Case Study of the U.S. VOS." *Renewable and Sustainable Energy Reviews* 137:110599; NESP. 2022b. "The National Standard Practice Manual (NSPM)." https://www.nationalenergyscreeningproject.org/national-standard-practice-manual; SMUD and E3. 2020. *Value of Solar and Solar + Storage*. Sacramento, CA: Sacramento Municipal Utility District and E3. https://www.smud.org/-/media/Documents/Rate-Information/2022-2023-Rate-Action/VOSstudy.ashx.

benefits and costs. The three main categories of benefits and costs discussed here are electricity system impacts, environmental and other externalities, and broader economy-wide effects.

Electricity System Impacts

The first set of benefits and costs discussed are those that occur within the boundaries of the electricity system. In many contexts, these benefits may be referred to as "avoided costs" of distribution, transmission, and generation though DG may also create additional costs (e.g., related to grid integration). A full accounting of electricity system impacts includes both the avoided and additional costs associated with DG deployment. The electricity system impacts discussed below include the effects of DG on energy generation costs, generation capacity costs, transmission and distribution capacity costs, and line losses, among other costs.

Energy generation costs: DG displaces electricity generation from other sources, thereby avoiding the costs related to the production of that electricity (primarily fuel and variable operating and maintenance costs). The specific electricity generation displaced by DG is a function of when the DG operates, where it is located, the quantity installed, and the period of the analysis (i.e., short-run or long-run).

In the short run, displaced generation is based on the existing fleet of generation resources in a region that serves the customers participating in net metering. In regions with organized wholesale power markets and with relatively low levels of DG penetration, the energy generation costs that DG avoids over the short run can be determined using hourly wholesale energy market prices. This wholesale energy market price generally represents the operating cost of the most expensive generating resource serving at any given time, which is the resource that DG would partially or fully displace.[29] In regions without organized wholesale power markets, or where DG penetration is large enough to impact the set of plants operating on the margin, some type of simulation is often required in order to predict what generation resources would have been operating but for the DG resources, in each hour of the year, and to then estimate the costs of producing electricity from these resource types.

Over the longer-term, displaced generation may need to be based on some set of projections or assumptions about the future resource mix in the study region, taking into

[29] During periods of scarcity or when imperfect competition exists, wholesale energy prices may rise above marginal operating costs.

consideration key economic and policy drivers for new generation build-out and retirement as well as underlying load growth.

Generation capacity costs: Maintaining a reliable grid requires sufficient electricity generation and demand response capacity at every point in time to meet customer and system needs.[30] To the extent DG reduces the need for generation capacity to maintain system reliability, it avoids the cost of building or contracting for that capacity. This avoided capacity benefit is again a function of the timing of DG operation, the electric demand profile[31] of a utility's customers, and available alternative resources to meet this demand.

Traditionally, avoided generation capacity costs from DG are determined by estimating how much DG would reduce demand at the time of system peak demand. However, as renewable penetration increases, capacity may be more valuable at times other than when system peak demand occurs. For example, in California and other areas with relatively high solar penetration, peak demand may still occur in the afternoon, but capacity is often most needed in the evening when solar starts to go offline but electricity demand does not decrease commensurately.[32] For this reason, the capacity value of solar generation (whether DG or utility-scale) tends to decline as solar penetration in a region increases. The intermittency of solar generation due to cloud cover may also reduce its capacity value. Pairing solar with storage, as has become common in both utility-scale and distributed applications, can help to mitigate both of those issues and enhance the capacity value of the combined resource.

The avoided capacity benefit that accrues to DG is equal to the amount of capacity that DG can defer multiplied by the per-unit cost of that deferred capacity. In regions with organized forward capacity markets, prices observed in those markets can be used to define the per-unit cost of avoided capacity. Otherwise, marginal capacity resources can be identified either through power sector modeling to determine what resources need

[30] This reliable capacity amount is usually estimated through stochastic models that determine the amount of capacity needed to maintain system-wide reliability under numerous electric demand and supply scenarios. Junlakarn, S., R. Diewvilai, and K. Audomvongseree. 2022. "Stochastic Modeling of Renewable Energy Sources for Capacity Credit Evaluation." *Energies* 15(14):5103.

[31] The electric demand profile is the aggregate amount of electric power demanded by all end users in a utility's territory over time.

[32] This phenomenon is popularly referred to as the duck curve because of the shape formed by gross and net-of-solar demand curves across a day. See Denholm, P., M. O'Connell, G. Brinkman, and J. Jorgenson. 2015. *Overgeneration from Solar Energy in California: A Field Guide to the Duck Chart.* Golden, CO: National Renewable Energy Laboratory. https://www.nrel.gov/docs/fy16osti/65453.pdf. Due to capacity constraints moving later in the day, the addition of BTM solar capacity with similar output and availability profiles diminishes each MW's ability to avoid generation capacity.

to be built for capacity, or as is commonly practiced, through less involved methods such as selecting lowest cost capacity resources based on recent market data.[33]

Transmission capacity costs: Transmission buildout is driven partly by asset age and condition, and partly by capacity needs (i.e., ensuring that electricity customers can be reliably served at all times). Increasingly, a significant amount of planned transmission capacity expansion is being driven by the need to access and integrate remote renewable generation.[34]

DG may defer the need for transmission upgrades if it can defer or avoid construction of specific transmission projects. However, unless DGs are planned and implemented to completely meet these location-specific capacity clean energy needs, or both, they may not eliminate the need for individual transmission projects.[35]

Even where DG is installed in an unplanned manner, it may still reduce overall demand in the system in a manner that could defer long-term transmission buildout. Some jurisdictions, like California,[36] account for the possibility of this benefit by assigning a long-run transmission investment reduction benefit for unplanned DG adoption by calculating transmission upgrade costs per unit increase in peak demand through either historic data or utility investment projections; these per-unit transmission ($/MW) upgrade costs

[33] The most common method of calculating avoided capacity cost is based on the net cost of new-entry (Net-CONE). Net-CONE is determined by first identifying the most likely, oftentimes lowest cost, capacity resource that would be procured to meet system needs and then subtracting the value of non-capacity benefits it provides the system (such as energy and ancillary services). Although natural gas combustion turbines have traditionally been the lowest net cost capacity resource in most of the country, decreasing battery storage costs combined with clean energy laws in many jurisdictions (which may make it unlikely for new gas peaker plants to be built) are changing this landscape. See SMUD (Sacramento Municipal Utility District) and E3. 2020. *Value of Solar and Solar + Storage.* Sacramento, CA: Sacramento Municipal Utility District and E3. https://www.smud.org/-/media/Documents/Rate-Information/2022-2023-Rate-Action/VOSstudy.ashx, p. 23. As the number of renewables on the grid increases, identifying marginal capacity resources using simplified methods may lead to skewed results, because both outstanding capacity needs and identifying the resource that can meet system capacity needs most cost-effectively depends on the interaction between regional electricity demand and renewable production characteristics.

[34] Joskow, P.L. 2020. "Transmission Capacity Expansion Is Needed to Decarbonize the Electricity Sector Efficiently." *Joule* 4(1):1–3.

[35] This brief compiles examples of non-wires-alternative projects implemented in specific locations to defer specific investments: Chew, B., E.H. Myers, T. Adolf, and E. Thomas. 2018. "Non-Wires Alternatives, Case Studies from Leading U.S. Projects." Smart Electric Power Alliance, Peak Load Management Alliance, and E4TheFuture. https://e4thefuture.org/wp-content/uploads/2018/11/2018-Non-Wires-Alternatives-Report_FINAL.pdf.

[36] E3. 2022. "2022 Distributed Energy Resources Avoided Cost Calculator Documentation: For the California Public Utilities Commission." https://www.cpuc.ca.gov/-/media/cpuc-website/divisions/energy-division/documents/demand-side-management/acc-models-latest-version/2022-acc-documentation-v1a.pdf, pp. 42–45.

are then assigned to peak hours that drive the need for incremental transmission capacity to serve electric demand. DG that saves or generates electricity during these peak demand times is assigned a proportional amount of avoided transmission benefit. Other jurisdictions, like Georgia,[37] currently attribute no avoided transmission cost to DG solar.

Regions with ambitious clean energy goals will likely need additional transmission to bring large amounts of renewables online economically. However, not many jurisdictions currently attempt to determine whether DG has another incremental transmission benefit that would accrue from avoiding transmission investments that could be needed to meet clean energy policy goals. One way to determine this benefit is to model and compare the transmission buildout necessary to meet future electricity demand and clean energy goals with and without DG. The difference in transmission buildout in that instance would provide an estimate of the impact of DG. Because transmission needs are location and time specific, however, the type, amount, and quantity of DG resources will drive the results of this analysis. This is a complex system analysis task. Sacramento Municipal Utility District's (SMUD's) recent value of solar and storage study simplified this by estimating the transmission upgrade costs necessary to bring more supply side solar projects online and then allocated this avoided cost to distributed solar.[38]

Distribution and sub transmission capacity and other costs: Like transmission upgrade and buildout, distribution system upgrades primarily are necessitated by either peak demand growth[39] or because equipment is at the end of its useful life. DG can avoid distribution costs in two ways.

First, DG can defer the need for distribution upgrades to meet growing electricity demand by reducing peak demand growth on distribution circuits. To realize this benefit, DG needs to consistently reduce demand at the right time and locations, for example, at circuits that are near capacity. As most DG is adopted and operated by customers, it can be hard to apply DG to defer distribution capacity needs in specific locations because a

[37] The Georgia Public Service Commission has adopted the following "framework" regarding the costs and benefits of solar resources: "During the winter, the system peaks generally occur just before or at dawn and after dusk. For these reasons, solar generation will not have an appreciable impact on deferring future transmission capacity investments." GPSC (Georgia Public Service Commission). 2017. "A Framework for Determining the Costs and Benefits of Renewable Resources in Georgia." GPSC Docket No. 40161, Document Filing #167588. https://psc.ga.gov/search/facts-document/?documentId=167588, p. 17.

[38] SMUD (Sacramento Municipal Utility District) and E3. 2020. *Value of Solar and Solar + Storage*. Sacramento, CA: Sacramento Municipal Utility District and E3. https://www.smud.org/-/media/Documents/Rate-Information/2022-2023-Rate-Action/VOSstudy.ashx, pp. 32–33.

[39] Peak demand growth in a specific distribution feeder will impact that specific feeder only. However, data to conduct such disaggregated analysis are hard to obtain and the analysis is itself complex. Simplifying assumptions are often applied to understand the aggregate probability of distribution upgrades needed due to aggregate growth in peak demand.

predetermined quantity of peak-load-reducing DG (e.g., community solar with storage) needs to be installed at that circuit once a need for an upgrade is identified and that DG needs to be installed and operational before the circuit gets overloaded. New York State Energy Research and Development Authority's (NYSERDA's) solar value stack[40] includes a locational system relief value that applies to specific parts of the grid; the California Public Utilities Commission's (CPUC's) distribution deferral analysis[41] identifies distribution upgrades that can be deferred by planned DGs.

California also accounts for the possible benefit that, in the long-run, unplanned customer-adopted DG has the aggregate effect of deferring distribution upgrades that otherwise would have been needed.[42] To account for these nonspecific deferrals, analysts can develop a correlation between peak load and distribution upgrade costs and then assign an on-average distribution system avoided cost to DG based on how much peak demand they reduce.[43]

Second, to the extent wear and tear of distribution equipment is a function of current flowing through the equipment, reducing electricity flow through these circuits permanently means that the equipment will likely last longer than if DG were not installed. This second impact is smaller because its value is the difference in present value of savings realized from deferring a project further out in time. However, growth of solar and other DG for which a utility has not planned can also increase the need for distribution upgrades and thus impose additional costs on utility customers.[44]

Transmission and distribution line losses: When DG reduces the amount of electricity that travels through transmission and distribution lines, it also reduces the amount of electricity lost, in the form of heat, in delivering that electricity from supply-side generation. These line losses should be accounted for when determining how much energy and peak

---

[40] NYSERDA. 2022c. "The Value Stack." https://www.nyserda.ny.gov/All-Programs/ny-sun/contractors/value-of-distributed-energy-resources.

[41] See California Public Utilities Commission's (CPUC's) Distribution Planning and Distribution Investment Deferral Framework here: CPUC. 2021a. "Distribution Planning." https://www.cpuc.ca.gov/industries-and-topics/electrical-energy/infrastructure/distribution-planning.

[42] E3. 2022. "2022 Distributed Energy Resources Avoided Cost Calculator Documentation: For the California Public Utilities Commission." https://www.cpuc.ca.gov/-/media/cpuc-website/divisions/energy-division/documents/demand-side-management/acc-models-latest-version/2022-acc-documentation-v1a.pdf, pp. 49–54.

[43] Care should be taken when combining the benefits of on-average distribution system avoided costs with specific distribution system deferral avoided costs, to avoid both double counting and undercounting the aggregate distribution system benefits that accrue to DGs.

[44] Recent research conducted at Stanford attributed two-thirds of the average distribution network price increase between 2003 and 2016 in California to required distribution system upgrades due to adoption of unplanned distributed solar by customers. See Wolak, F.A. 2018. *The Evidence from California on the Economic Impact of Inefficient Distribution Network Pricing*. Cambridge, MA: National Bureau of Economic Research. https://www.nber.org/papers/w25087.

capacity needs DG can avoid. The total energy DG avoids is equal to the total energy saved or generated onsite, plus the amount of electricity that would have been lost in the form of heat through the transmission and distribution system.[45] Transmission and distribution line losses can either be calculated through detailed power flow modeling or via simplified engineering estimates.[46]

Other electricity system impacts: DG can also provide other electricity system benefits like ancillary services and financial risk reduction. "Ancillary services" is a catch-all term for services, in addition to electricity generation and capacity, that generation and storage resources provide the grid to maintain stability. These include voltage response, frequency response, spinning reserves, and operating reserves. Not all DG provide all these ancillary services; for example, dispatchable storage can provide frequency response, but distributed solar by itself typically cannot.

Alternatively, DG can also cause overloading of circuits (e.g., with distributed solar that could increase the need for fast frequency response and reserves on those circuits). Because the magnitude of this ancillary service avoided cost is small, this avoided cost is generally estimated by studying historic values of ancillary services and then projecting them forward or through detailed production cost modeling that estimates the value of future ancillary services.[47]

To the extent a utility replaces exposure to the wholesale gas market with customer-adopted DG, it reduces the risk of future fuel price volatility. This avoided cost is a function

[45] Line losses are a function of the square of the power flowing through the wire, the total resistance of the wire, and inversely proportional to the square of the voltage flowing through the wire. So, increases in voltage for a given quantity of power can reduce the line loss (note—this is how high voltage transmission minimizes line losses). During times of peak demand the amount of power flowing through the grid is relatively higher than other times; line losses avoided by DGs are thus greater during times of greater demand.

[46] Many jurisdictions use standard or average line loss factors to estimate line losses avoided by DG; these are then applied to the DG load-shape (i.e., the timing of generation) to determine the total amount of electricity and peak demand purchases from supply side resources the utility would otherwise have to make. For example, the SMUD value of solar study applies line loss factors ranging between 2.5 percent and 6 percent to determine the gross energy and capacity impact of DGs (SMUD and E3 2020, pp. 34–46). The Georgia Framework (GPSC 2017, p. 17) specifies that "Transmission energy losses should be included as a benefit and should be included through use of the projected Avoided Energy Costs. Transmission capacity losses should be included as a benefit for distributed resources"; however, transmission losses are described as difficult to value, and no methodology is prescribed.

[47] For example, the Sacramento Municipal Utilities Department Value of Solar study found that avoided costs for all ancillary services are at most $0.65 per MWh (which is a fraction of a percent of retail rates). To the committee's knowledge, studies do not estimate the value of individual avoided ancillary service costs from DG. For additional information on ancillary business models, see Wu, K., Y.M. Zhang, R. Hardison, and E. Weber. 2021. *Business Models to Accelerate the Utilization of Distributed Energy Resources*. NREL/TP-6A20-79549. Golden, CO: National Renewable Energy Laboratory. https://www.nrel.gov/docs/fy22osti/79549.pdf.

of the amount of exposure a utility has to fuel price volatility, the load shape and pattern of production of each DG, and the probability that the DG will reliably produce electricity or reduce load through its useful life.

Finally, there are other technological costs and benefits of integrating wide-spread DG adoption. As explained further in Chapters 5 and 6, DG can provide both individual and grid-level resilience benefits, and its grid resilience and economic value can be maximized through infrastructure that can help control DG operation. However, this infrastructure, including communications and cybersecurity, will require distribution investment with associated costs.

Environmental and Other Externalities

When DG displaces fossil fuel power plant operation, it can reduce both carbon emissions and pollution that would have occurred if the fossil power plant would have operated, as well as upstream environmental impacts of fossil fuel production and use. To the extent DG mitigates the build out of large-scale renewables, fossil power plants, and associated transmission (as discussed above), it can also reduce any associated environmental harm. These are among the societal environmental externalities that can accrue to a DG resource.

The price of electricity generally does not account for this societal cost or externality of greenhouse gas emissions or other pollutants from fossil-fuel power plants. Some states, like California, Connecticut, Delaware, Maine, Maryland, Massachusetts, New Hampshire, New Jersey, New York, Oregon, Rhode Island, Vermont, Virginia, and Washington have clean energy goals and/or carbon cap-and-trade programs that directly or indirectly internalize some of the cost of carbon emissions.[48] Whether and to what extent to include externalities in utility avoided costs is a policy decision based on the laws and policy priorities of the jurisdiction in question. How to internalize these externalities and include them in avoided costs is an economic question that regulators are tasked with resolving.[49]

DG and supply side clean energy resources can avoid air pollution and carbon emissions. The avoided costs of these externalities should be treated consistently. Inconsistency in

---

[48] Center for Climate and Energy Solutions. 2023. "U.S. State Carbon Pricing Policies." https://www.c2es.org/document/us-state-carbon-pricing-policies; NCSL. 2021. "State Renewable Portfolio Standards and Goals." https://www.ncsl.org/energy/state-renewable-portfolio-standards-and-goals.

[49] The National Standard Practice Manual, for example, explains how to value carbon when assessing DER. NESP. 2022b. "The National Standard Practice Manual (NSPM)." https://www.nationalenergyscreening-project.org/national-standard-practice-manual.

how these externalities are valued among resources could lead to inefficient investment and thus raise electricity prices more than necessary.

Other externalities such as the monetary value of reduced land use and private customer resilience[50] may exist as well. These values have not yet been well studied, and policymakers should consider whether and to what extent these externalities should be accounted for when developing net metering policies and compensation levels.

Greenhouse gas (GHG) reduction: This section explains one major category of externalities avoided by DG, carbon and air pollution, and the different approaches to quantify and monetize these externalities. The avoided social cost of GHG emissions from clean electricity consumption is equal to the product of the total GHG emissions abated by the DG and an estimate of the cost of carbon (further explained below). The total carbon emissions abated by DG is a product of the load shape of the DG times the marginal carbon emissions of the electricity grid through the useful lifetime of the DG.[51]

The marginal carbon dioxide (CO_2) emissions from a grid depend on the resource mix of the grid.[52] A grid with only fossil fuel resources will have predictable marginal emissions factors, generally equal to the emissions factors of gas-fired peaker plants during peak hours and equal to the emissions from load-following plants at other times. The marginal emissions of a grid with an increasing number of renewables will vary significantly with time; they will also evolve over time. When renewables are on the margin, DG will not abate carbon emissions; when a mix of renewables and fossil-fired generation is on the margin, then the amount of CO_2 emissions DG will abate will be greater than zero, but less than if only fossil-fired generation was on the margin.

[50] Resilience from combined solar and storage systems is an important benefit for customers dependent on electricity to power medical equipment and other vulnerable populations, as well as any customers facing frequent electricity outages. As currently structured, net metering is too broad a tool to ensure that these specific customers receive the resilience benefits of DG. Policymakers may want to consider alternative solutions for providing solar and storage to these customers, particularly reducing the barrier posed by their installation costs (see Chapter 5).

[51] The load shape of the DG, or its time-varying electricity production, is generally estimated through engineering calculations or through measurement and verification, or both. For example, the electricity generation shape of distributed rooftop solar is a function of the solar irradiation on the solar panel and can be easily estimated through standard tools once the properties of a solar panel are known. The load shape of combined solar and storage installation is more uncertain as it depends on the operating profile of the storage system, which can be programmed by the end-user. See, for example, PVWatts® by the National Renewable Energy Laboratory: NREL. 2020. "PVWatts® Calculator." https://pvwatts.nrel.gov.

[52] Holland, S.P., M.J. Kotchen, E.T. Mansur, and A.J. Yates. 2022. "Why Marginal CO_2 Emissions Are Not Decreasing for US Electricity: Estimates and Implications for Climate Policy." *Proceedings of the National Academy of Sciences* 119(8):e2116632119.

The committee believes that an estimate of the marginal emissions for the very near-term can be developed through wholesale energy market data where available.[53] Estimates of future emissions abated by DG require an estimate of future grid operations as explained above. There are two main methods to set a price of carbon emissions: the damage and social cost of carbon and the marginal abatement cost of carbon.

The Social Cost of Carbon (SCC) is an estimate of the net economic cost of future damages that would be caused by incremental CO_2 emissions.[54] The SCC is calculated using a combination of complex global models that collectively forecast and quantify future damages due to increased temperature and then monetize these damages.[55] These models produce a range of possible future damages. The present value of this range is dependent on the discount rate being applied. The Interagency Working Group's 2016 Technical Support Document[56] aggregated the range of SCCs.[57]

The Marginal Abatement Cost (MAC) is the cost of reducing the last ton of carbon to comply with a predetermined carbon reduction target. For example, California has a mandate for the electricity sector, per Senate Bill 100, to get to zero carbon electricity sales by 2045.[58] The California Public Utilities Commission (CPUC) conducts capacity expansion modeling[59] to determine the total cost and the marginal cost of meeting this

[53] Ideally these emissions estimate would vary by location to accurately represent the emissions factors of power plants displaced. See, for example, work done by PJM on locational marginal emissions: PJM. 2022. "Locational Marginal Emissions Introduction." https://www.pjm.com/-/media/committees-groups/forums/emerging-tech/2022/20220614/item-3---locational-marginal-emissions-introduction.ashx.

[54] For more information on the social cost of carbon, see NASEM. 2017. *Valuing Climate Damages: Updating Estimation of the Social Cost of Carbon Dioxide*. Washington, DC: The National Academies Press.

[55] Although there are many damages hard to quantify and monetize, such as increased likelihood of wildfires and flooding that these models either omit or incompletely account for (see Table 2, p. 4 in Institute of Policy Integrity. 2017. *The Social Cost of Greenhouse Gases and State Policy*. New York: Institute for Policy Integrity), and future damage valuations are uncertain, these are the best available estimates of the global social cost of carbon emissions.

[56] Interagency Working Group on Social Cost of Greenhouse Gases, United States Government. 2016. "Technical Support Document: Technical Update of the Social Cost of Carbon for Regulatory Impact Analysis—Under Executive Order 12866." https://www.epa.gov/sites/default/files/2016-12/documents/sc_co2_tsd_august_2016.pdf, p. 4.

[57] These aggregated values range from $14 to $148 in 2020 per metric ton of carbon dioxide to $31 to $254 per ton in 2050, with a mean value of $50 in 2020 and $83 per ton in 2050. See Paul, I., P. Howard, and J.A. Schwartz. 2017. *The Social Cost of Greenhouse Gases and State Policy*. New York: Institute for Policy Integrity, New York University School of Law. https://policyintegrity.org/files/publications/SCC_State_Guidance.pdf.

[58] For more information on Senate Bill (SB) 100, see California Energy Commission. 2023. "SB 100 Joint Agency Report." https://www.energy.ca.gov/sb100.

[59] See CPUC avoided cost calculator documentation at pp. 30–31: E3. 2022. "2022 Distributed Energy Resources Avoided Cost Calculator Documentation: For the California Public Utilities Commission." https://www.cpuc.ca.gov/-/media/cpuc-website/divisions/energy-division/documents/demand-side-management/acc-models-latest-version/2022-acc-documentation-v1a.pdf.

target to the electricity sector in each year and applies the estimate of marginal cost, also called the shadow price of carbon, to develop a cost of clean energy compliance, called the GHG Adder, that is used as an avoided cost for all DG.

For regions with binding carbon reduction targets, the MAC represents the price on carbon that will help attain binding carbon reduction targets most cost-effectively. Accordingly, regulators and/or utilities should internalize the MAC into avoided costs.[60] For regions without binding targets, the value of carbon included in utility avoided costs is a policy choice. The SCC, which is a global measure of carbon damage, and estimates of MAC from different levels of carbon reduction, can help inform this decision.

Air pollution: The existing electric grid in the United States relies, in part, on combustion of gas, coal, oil, and biomass resources which can produce health-damaging air pollutants in addition to greenhouse gasses. These pollutants include particulate matter ($PM_{2.5}$ and PM_{10}), nitrogen oxides (NO_x), sulfur dioxide (SO_2), volatile organic compounds (VOCs), and heavy metals among others. NO_x and SO_2 further react in the atmosphere to produce more $PM_{2.5}$ and ozone. Exposure to air pollution, especially $PM_{2.5}$, is associated with cardiovascular and respiratory health impacts for exposed populations. These include, but are not limited to, asthma exacerbations, adverse birth outcomes, heart attacks, and increased mortality.[61]

The distribution of public health impacts of power plant emissions—or the benefit of a reduction in emissions—depends on the location of each power plant, the magnitude of geographic area affected by the emitted pollutants, and characteristics of the populations exposed to the pollutants (discussed further in Chapter 5).

To determine this health impact, analysts multiply the energy profile of DG with an emissions factor that represents the expected density of pollutants from the generation mix

---

[60] The MAC applied to DG should (1) be aligned with the spending the utility would have to make to meet their carbon goals if customers did not adopt DGs going forward, and thus (2) put DGs on an even playing field with supply side renewables which the utility would purchase absent DGs.

[61] The vast majority of the monetary valuation of health impacts from criteria air pollution exposure are a result of PM2.5 caused mortality risk (Massetti, E., M. Brown, M. Lapsa, I. Sharma, J. Bradbury, C. Cunliff, and Y. Li. 2017. *Environmental Quality and the U.S. Power Sector: Air Quality, Water Quality, Land Use and Environmental Justice*. ORNL/SPR-2016/772. Oak Ridge, TN: Oak Ridge National Laboratory. http://info.ornl.gov/sites/publications/files/Pub60561.pdf). See also Fann, N., A.D. Lamson, S.C. Anenberg, K. Wesson, D. Risley, and B.J. Hubbell. 2012. "Estimating the National Public Health Burden Associated with Exposure to Ambient PM2.5 and Ozone." *Risk Analysis: An International Journal* 32(1):81–95.

on a regional grid that DG will replace.[62] The difference in emissions between the two scenarios is a function of the quantity, load shape, and location of the DG installed. Atmospheric dispersion modeling of these emissions can then help evaluate air pollutant concentrations where people live under each scenario. This can then be coupled with concentration-response models to estimate the epidemiological impacts of air pollutant emissions reductions associated with DG.[63] Some jurisdictions simplify this analysis by first calculating the average hourly emissions of the electric grid being evaluated. The aggregate load shape of the DG in question is then applied to these hourly emissions factors to determine the total amount of emissions abated by the DGs. Care should be taken when adding the avoided costs of health damage to avoided costs of carbon reduction to make sure that these two avoided costs are appropriately additive and are not being double counted.[64]

[62] In its most simplistic form, this emissions factor can be estimated by developing a grid average marginal emissions factor using the makeup of the resources that participate in the wholesale market. A more detailed and relatively more accurate set of emissions factors can be determined through a detailed production cost model of a regional grid both with and without the DG.

[63] Certain models, such as the U.S. Environmental Protection Agency's (EPA's) Co-Benefits Risk Assessment Model (COBRA), then assign a monetary value to these modeled health outcomes using such measures as the Value of a Statistical Life. See EPA. 2023. "CO-Benefits Risk Assessment Health Impacts Screening and Mapping Tool (COBRA)." https://www.epa.gov/cobra. To reduce the computational costs of air dispersion modeling, standardized emission factors and benefits are sometimes applied to avoided emissions. For example, EPA has standardized monetized values for every ton of avoided emissions for $PM_{2.5}$ and $PM_{2.5}$ precursors: EPA (U.S. Environmental Protection Agency). 2022. "Sector-Based PM2.5 Benefit per Ton Estimates." https://www.epa.gov/benmap/sector-based-pm25-benefit-ton-estimates. Studies relying on these methods have been used to estimate human health benefits from DG or other clean energy resources, for example: Mantegna, G., A. Burdick, S. Price, A. Olson, M.M. Kinnon, and S. Samuelsen. 2021. *Quantifying the Air Quality Impacts of Decarbonization and Distributed Energy Programs in California*. San Francisco, CA: E3, and Irvine, CA: Advanced Power and Energy Program. https://www.ethree.com/wp-content/uploads/2022/01/CPUC-Air-Quality-Report-FINAL.pdf.

[64] If a utility takes a SCC approach to monetizing carbon, then taking a damage cost approach to avoided health impacts and including them as an avoided cost for DG would be appropriate as long as these externalities are consistently included in payments for all clean energy resources that beget the same benefits. MAC estimates also account for the costs of abating pollutants of fossil power plant operation. For example, in California, utilities are required to contract with carbon-free resources and procure a certain amount of DER to meet California's clean energy mandates and the state's carbon dioxide reduction goals. These clean energy contracts displace power plant operation. When power plant operation is displaced, carbon dioxide emission, other air pollutants, and methane leakage are reduced; they all represent avoided costs attributable to displacement of power plant operation. Therefore, the MAC also accounts for the abatement of these pollutants.

Economy-wide Effects

In addition to the direct effects on the electricity system and environmental externalities, net metering and other policies that encourage DG deployment may lead to broader economy-wide effects. These effects are often challenging to precisely quantify in monetary terms, and as such are often left outside the scope of formal benefit-cost analysis but may nevertheless be relevant to policy discussions around the merits of net metering and other DG deployment policies.

The first of these broader economy-wide implications is the potential for helping to drive down the costs of certain DG resources by accelerating their deployment. This effect is sometimes described as the "learning curve," and has been a key rationale for net metering and other policies aimed at supporting the deployment of clean energy technologies in the early stages of market development, before the technologies and their associated value chain businesses have reached some threshold scale.[65] For rooftop solar, a portion of the overall cost of an installed system is associated with hardware components that are procured in global markets, and their cost trajectories reflect growth in both global supply and demand. Other costs, such as installation labor, reflect local conditions, which can be more readily influenced through local policies, such as net metering. Another potential economy-wide effect of net metering is its impact on energy security. In the case of rooftop solar, the underlying fuel source (the sun) is, in some sense, a domestic source, requiring no foreign fuel sources.

Finally, net metering and other DG support policies have sometimes been advanced for their employment and workforce development impacts. According to the latest Solar Jobs Census, roughly 135,000 individuals were employed in residential and commercial solar installation and development in 2021, representing roughly 50 percent of all domestic jobs in the U.S. solar industry, which in turn represent about 40 percent of all jobs in the U.S. electric power sector.[66] Net metering and other forms of DG policy support have contributed to employment growth in the U.S. solar industry. Assessing the net effects on jobs requires consideration of jobs lost or displaced in other energy sectors; indirect effects associated with changes in energy prices; and job quality (e.g., pay, benefits, safety, and security), as well as concerns about distributional impacts—jobs gained and lost may occur in different locations or among different populations. These distributional issues are discussed further in Chapter 5.

[65] See, for example, van Benthem, A., K. Gillingham, and J. Sweeney. 2008. "Learning-by-Doing and the Optimal Solar Policy in California." *The Energy Journal* 29(3):131–151.

[66] IREC. 2019. "Solar Jobs Census 2019." https://irecusa.org/resources/national-solar-jobs-census-2019.

FINDINGS

Finding 4-1: Net metering can provide a compelling financial incentive for customers to install DG in areas with relatively high retail electricity rates, while in regions with low electricity prices, it may be insufficient on its own to motivate significant levels of adoption.

Finding 4-2: From the perspective of society overall, whether net metering provides an economically efficient level of compensation for DG depends on how the volumetric energy portion of the customer's retail electricity tariff compares to the marginal cost of producing and delivering electricity, including environmental externalities, at the time and location where DG production occurs.

Finding 4-3: Both volumetric electricity prices and the marginal cost of producing and delivering electricity can, and do, vary significantly from one situation to another, reflecting broad regional differences in the generation resource mix and policies, localized factors, and the temporal profile of individual DG resources. As a result, in some regions net metering may overcompensate DG, while in other regions it may under-compensate DG, relative to an economically efficient level.

Finding 4-4: The social marginal costs of generating and delivering electricity, including externalities, are likely to change significantly over time, particularly with modifications in the generation mix and its cost characteristics. Any assessment of the economic merits of net metering, and its potential rate impacts to non-participants, must consider these dynamic factors.

Finding 4-5: The major components of social marginal costs of generating and delivering electricity are the avoided costs of electric generation, capacity, deferred transmission and distribution infrastructure, avoided line losses, and decreases in greenhouse gases and air pollution resulting from reduced fossil plant operation.

Finding 4-6: Utilities and regulators can develop more economically efficient DG compensation, either by modifying the retail rate structures so that volumetric energy charges better align with social marginal costs or by modifying the net metering mechanism to transition to net billing or buy-all and sell-all tariffs with compensation levels reflective of social marginal costs.

Finding 4-7: Setting DG compensation at a level reflective of social marginal costs may induce higher or lower levels of DG deployment than would occur under a traditional

NEM structure with volumetric rates that recover both fixed and variable costs. The direction of the impact depends on the level of social marginal costs, which vary by location and point in time, and how those compare to volumetric retail rates.

Finding 4-8: The absolute magnitude of any rate impacts associated with net metering scale with the penetration of DG resources. As shown earlier in Chapter 3, in most U.S. states, less than 1 percent of residential customers are currently taking service under a net metering tariff. Where this is the case, rate impacts are minimal.

Finding 4-9: The rate impacts of net metering on non-participating customers depend on how the cost impacts to the utility compare to the revenue impacts. When net metering is coupled with retail rate structures that recover fixed utility costs via the volumetric energy (cents/kWh) component, revenue reductions will generally exceed utility cost reductions, leading to upward pressure on non-participant rates for the utility to recover its costs.

Finding 4-10: When net-metered DG displaces fossil fuel power plant operation, environmental benefits are realized due to decreased greenhouse gas emissions and air pollution. The benefits of reduced carbon emissions accrue globally, while the health benefits of reduced air pollution are region specific as they accrue to customers downwind of the displaced output from affected power plants.

Finding 4-11: Environmental benefits achieved by net-metered DG may also be accomplished through other policies and technologies, potentially at a lower cost to non-participating customers.

Finding 4-12: Utilities and regulators balance many competing objectives when setting rates and DG compensation structures. There may be scenarios where some level of non-participant rate impacts is deemed acceptable when it advances other ratemaking objectives (e.g., economic efficiency) or policy goals.

RECOMMENDATIONS

Recommendation 4-1: Regulators should strive to develop retail rate structures—for both DG and non-DG customers—with usage-based energy prices that correspond as closely as possible to the social marginal cost of producing and delivering electricity, inclusive of time- and location-specific environmental or other externality costs (insofar as they are not already internalized), while recognizing

other competing rate-design objectives. Under this condition, net metering will generally result in economically efficient levels of investment in DG technologies, albeit with some rate impacts to non-participating customers.

Recommendation 4-2: In the absence of economically efficient rate structures for all customers, the alternative solution may be to implement changes to the net metering mechanism—either buy-all or sell-all, or net billing tariffs—for DG customers, with DG compensation levels set at, or near, the social marginal cost of electricity production and delivery.

Recommendation 4-3: When evaluating the economic implications of current or proposed changes to net metering rules, decision makers should consider specific conditions in their jurisdiction (e.g., the generation mix and costs, DG penetration levels, clean energy policies, etc.), while also anticipating how those conditions may evolve over the lifetime of DG resources. Environmental benefits should be carefully weighed alongside any rate impacts to non-participating customers.

Recommendation 4-4: In some cases, additional compensation for DG customers may be warranted above and beyond what can be provided through net metering or other DG tariffs. In these circumstances, any additional incentives for DG customers should be provided through transparent incentives designed to achieve a specified outcome (e.g., to encourage adoption by low-income customers, to relieve constraints in specific localized regions of the grid, to meet legislative DG targets, etc.) and, if appropriate, funded from sources outside of electricity rates.

Recommendation 4-5: In determining DG compensation, policymakers should explicitly account for the environmental and other externalities that DG reduces when it displaces fossil fuel generation. To fairly compensate DG for this clean energy attribute, policymakers should consider whether to do so based on the monetary value of the environmental externalities that the DG is likely to abate or the cost to achieve the same reduction in environmental harms through alternative technologies or policies.

Recommendation 4-6: Utilities should work with regulators and stakeholders to identify the types and formats of information (such as heat maps and avoided-cost calculations) that the utility should provide publicly about their distribution systems and their provision of electricity service so that suppliers and potential customers of behind-the-meter generation can see the benefits and costs that

behind-the-meter technologies can provide to the local electric system. Regulators and decision makers of publicly owned utilities should ensure that such information is provided to the public in easily understandable and accessible ways.

Recommendation 4-7: To the extent possible, BTM DG compensation rates should be location- and time-based, tied to available distribution capacity, to indicate where that DG would have the greatest value (e.g., in relieving or avoiding distribution constraints and/or displacing local generation with high emissions). Considering the locational value of solar can have particular value to low-income customers who often live in areas with emitting generation and poorer air quality.

Equity Considerations of Net Metering

INTRODUCTION

This chapter examines equity considerations related to net metering. Prior chapters have discussed net metering policies and their variants, trends in the costs of distributed generation, and economic considerations in the design of net metering and its variants. These economic considerations include a discussion of the different benefits and costs of behind-the-meter (BTM) distributed generation (DG) supported by net metering for customers who are able to take advantage of it and those who do not. Because low-income households, racial and ethnic minority populations, and renters are more likely to be non-participants, economic transfers and any differential net metering benefits and costs between participating and non-participating customers have equity implications.[1,2]

This chapter addresses the equity issues associated with the distribution of benefits and impacts of DG, the role of net metering in shaping this distribution, and net metering design considerations to address these inequities. The chapter begins with a discussion of the concept of equity, highlights the diverse characteristics of electricity customers and existing inequities in the electricity system and how such inequities intersect with net metering policy, and examines the direct, indirect, and societal impacts of net metering. This discussion includes the impacts of net metering on participants and non-participants and on different subgroups.

The specific equity-related topics examined in this chapter include:

- An overview of the multiple dimensions and types of equity in the context of net metering and distributed generation
- The demographics of net metering participation
- Equity concerns associated with net metering non-participants
- Environmental, health, and societal equity effects from net metering policies

[1] Sunter, D.A., S. Castellanos, and D.M. Kammen. 2019. "Disparities in Rooftop Photovoltaics Deployment in the United States by Race and Ethnicity." *Nature Sustainability* 2(1):71–76.

[2] Forrester, S., G. Barbose, E. O'Shaughnessy, N. Darghouth, and C.C. Montañés. 2022. *Residential Solar-Adopter Income and Demographic Trends: November 2022 Update*. Berkeley, CA: Lawrence Berkeley National Laboratory.

- Current state of equity requirements and range of equity solutions
- Findings and recommendations

Examination of equity considerations is part of a growing recognition that public investments in climate mitigation and adaptation, including support and incentives for renewable DG, have implications for improving, and avoiding worsening, outcomes related to a fair and just energy system. The 2021 National Academies *Accelerating Decarbonization of the U.S. Energy System* report emphasizes the opportunity to convert the U.S. fossil fuel economy into a green economy that is environmentally sustainable, economically secure, and socially just.[3] Advocates of this opportunity describe the need to compensate for historically inequitable energy policies and programs and to prioritize program participation in historically burdened populations that have been exposed to significant pollution from power plants and other fossil fuel infrastructure, as well as households and small businesses that face high energy cost burdens.

Residential decarbonization efforts have historically deployed numerous programs and policies that have been inaccessible to disadvantaged populations and environmental justice communities.[4] As a result, energy cost burdens remain persistently high for low-income households and people of color, households with children and elderly residents, and in the South and rural America, where housing is generally less energy-efficient.[5] Like many other public and ratepayer-funded clean energy initiatives, these same sociodemographic subgroups, and the South,[6,7] have been under-represented in net metering programs.

---

[3] National Academies of Sciences, Engineering, and Medicine (NASEM). 2021. *Accelerating Decarbonization of the U.S. Energy System*. Washington, DC: The National Academies Press. https://doi.org/10.17226/25932.

[4] U.S. Environmental Protection Agency's Climate and Economic Justice Screening Tool provides indexes and indicators that enable mapping of communities by the extent of their vulnerabilities and risks. The map uses publicly available, nationally consistent datasets. Learn more about the methodology and datasets that were used to identify disadvantaged communities in the current version of the map available at: Council on Environmental Quality. 2022. "Climate and Economic Justice Screening Tool." https://screeningtool.geoplatform.gov/en/#3/33.47/-97.5. See also Dolšak, N., and A. Prakash. 2022. "Three Faces of Climate Justice." *Annual Review of Political Science* 25:283–301.

[5] Brown, M.A., and M. Ahmadi. 2019. "Would a Green New Deal Add or Kill Jobs?" *Scientific American*. https://www.scientificamerican.com/article/would-a-green-new-deal-add-or-kill-jobs1; Brown, M.A., J. Hubbs, X.V. Gu, and M.-K. Cha. 2021. "Rooftop Solar for All: Closing the Gap Between the Technically Possible and the Achievable Potentials." *Energy Research and Social Science* 80(October):102203; Drehobl, A., L. Ross, and R. Ayala. 2020. *How High Are Household Energy Burdens? An Assessment of National and Metropolitan Energy Burden Across the United States*. Washington, DC: American Council for an Energy-Efficient Economy.

[6] Smith, K.M., C. Koski, and S. Siddiki. 2021. "Regulating Net Metering in the United States: A Landscape Overview of States' Net Metering Policies and Outcomes." *The Electricity Journal* 34(2):106901.

[7] Sunter, D.A., S. Castellanos, and D.M. Kammen. 2019. "Disparities in Rooftop Photovoltaics Deployment in the United States by Race and Ethnicity." *Nature Sustainability* 2(1):71–76.

Participants in net metering are more likely to be from affluent, well-educated, and/or White households.[8] Low levels of solar rooftop adoption and participation in net metering tariffs have been documented in numerous disadvantaged communities[9,10] and disadvantaged populations across multiple U.S. cities in regions around the nation.[11] Studies[12] indicate that some of the benefits of net metering could bypass some of the most disadvantaged and overburdened populations, as well as environmental justice and "frontline" communities, unless programs are designed to break down barriers to participation and reach out to disadvantaged and non-traditional populations. In addition, to the extent that there is cost-shifting from net metering participants to net metering non-participants in particular states and utility systems, less affluent customers and disadvantaged communities risk bearing a disproportionate share of the cost as they are more likely to be non-participants. If the environmental, grid security, resilience, and health benefits of BTM DG supported by net metering are considered, they may offset (or more than offset) any cost shift to non-adopters. However, the body of evidence is not sufficient to draw definitive conclusions, and the evaluation of such cost shifts is inherently location dependent.

Although historic adoption of DG has been primarily among well-educated, wealthy, and White households, recent initiatives are focusing on modifying programs so that they provide more benefits to disadvantaged and environmental justice (EJ) communities and evidence has shown a broadening of rooftop solar participation across income groups.[13] While in 2021, approximately 3 percent of solar adopters earned less than 120

[8] Forrester, S., G. Barbose, E. O'Shaughnessy, N. Darghouth, and C.C. Montañés. 2022. *Residential Solar-Adopter Income and Demographic Trends: November 2022 Update*. Berkeley, CA: Lawrence Berkeley National Laboratory; Wolske, K.S., P.C. Stern, and T. Dietz. 2017. "Explaining Interest in Adopting Residential Solar Photovoltaic Systems in the United States: Toward an Integration of Behavioral Theories." *Energy Research & Social Science* 25:134–151.

[9] Lukanov, B.R., and E.M. Krieger. 2019. "Distributed Solar and Environmental Justice: Exploring the Demographic and Socio-Economic Trends of Residential PV Adoption in California." *Energy Policy* 134:110935.

[10] Forrester, S., G. Barbose, E. O'Shaughnessy, N. Darghouth, and C.C. Montañés. 2022. *Residential Solar-Adopter Income and Demographic Trends: November 2022 Update*. Berkeley, CA: Lawrence Berkeley National Laboratory.

[11] Reames, T.G. 2020. "Distributional Disparities in Residential Rooftop Solar Potential and Penetration in Four Cities in the United States." *Energy Research & Social Science* 69:101612.

[12] Referenced in footnotes 3, 4, 5, and 6.

[13] Forrester, S., G. Barbose, E. O'Shaughnessy, N. Darghouth, and C.C. Montañés. 2022. *Residential Solar-Adopter Income and Demographic Trends: November 2022 Update*. Berkeley, CA: Lawrence Berkeley National Laboratory.

percent of their area's median income,[14] solar adoption among low- and moderate-income households has been expanding in recent years. Enablers of this democratization of rooftop solar include:

- Declining solar costs
- Improved affordability enabled by new financing techniques
- Opportunities for virtual net metering and community solar
- Local policies to address barriers to adoption
- The inclusion of solar as an eligible measure in the Department of Energy's (DOE's) Weatherization Assistance Program[15] as a result of the Energy Act of 2020
- The creation of a DOE Weatherization Readiness Fund[16] in 2022 that can be used to repair roofs, readying them for solar panels
- Components of the 2022 Inflation Reduction Act, including solar tax credits

These enablers are likely to continue to expand rooftop solar adoption by a broader range of households.

DEFINITION AND DIMENSIONS OF EQUITY

Definition and Scope of Equity Concerns

Equity is a multi-dimensional concept including distributional, procedural, intergenerational, and structural equity.[17] When applied to the net metering of DG, these concepts can be considered across various socio-demographic groups, timeframes, and geographic scales. From a systems approach, the equity impacts of net metering can be direct, indirect, or societal in nature.

[14] Ibid. See also DOE. 2021. "Solar Futures Study." https://www.energy.gov/eere/solar/solar-futures-study, p. 82.

[15] For more information about DOE's Weatherization Assistance Program (WAP) see https://www.energy.gov/scep/wap/weatherization-assistance-program.

[16] For more information about DOE's Weatherization Readiness Fund see https://www.energy.gov/scep/wap/articles/weatherization-program-notice-22-6-weatherization-assistance-program.

[17] Brown, M.A., A. Soni, M.V. Lapsa, K.A. Southworth, and M. Cox. 2020b. "High Energy Burden and Low-Income Energy Affordability: Conclusions from a Literature Review." *Progress in Energy* 2(4):1–35; Sovacool, B.K., and M.H. Dworkin. 2015. "Energy Justice: Conceptual Insights and Practical Applications." *Applied Energy* 142:435–444.

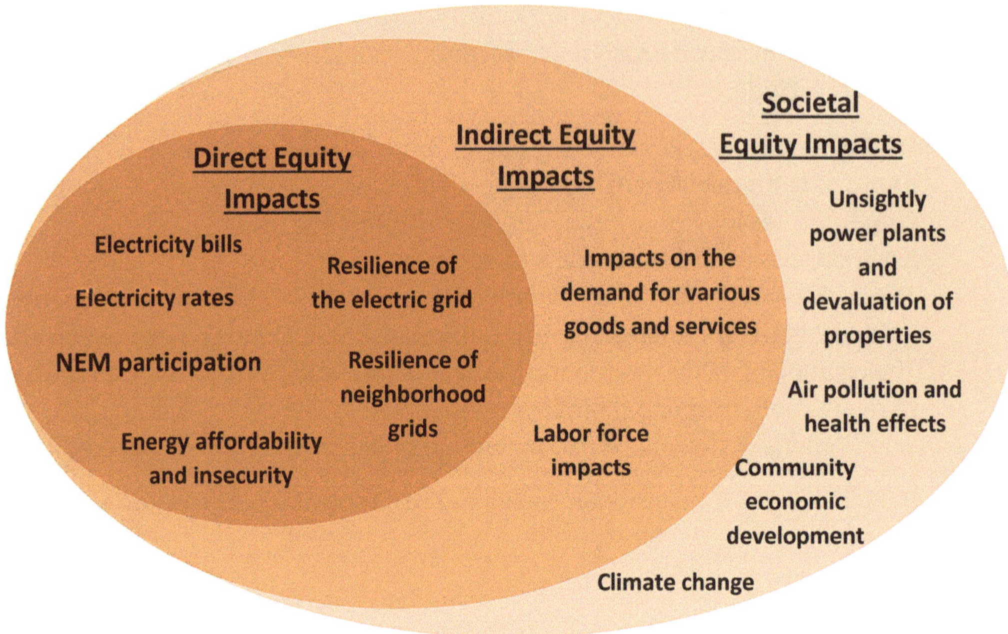

FIGURE 5-1 Direct, indirect, and societal impacts of net metering on equity.

Figure 5-1 illustrates how equity impacts can involve direct effects, such as impacts on electricity bills and rates and on the resilience of neighborhoods and the grid. They can also be indirect, as a function of job opportunities and impacts on the demand for various goods and services, which can lead to shifts in economic prosperity of neighborhoods and regions. Finally, there are societal equity effects, including public health, climate change, and community economic development, that can result in the decline or growth of communities and regions. The intricate web of linkages and effects will advantage some places over others in ways that can be difficult to foresee. As Dr. Martin Luther King, Jr., so eloquently summarized: "We are caught in an inescapable network of mutuality, tied in a single garment of destiny. Whatever affects one directly, affects all indirectly."[18]

Direct equity impacts of net metering include various first-order effects. Electricity rates and bills of participating customers would be directly impacted by net metering as their adoption of DG would be expected to reduce their electricity costs and provide bill

[18] King, Jr., M.L. 1963. "Letter from Birmingham Jail." April 1.

stability. The electricity rates and bills of non-participating customers may also be affected, depending on the penetration of DG, the compensation rates offered, and how the costs of net metering and its variants and alternatives are recovered (see Chapter 4). In addition, the expansion of DG may have resilience benefits or costs to net metering participants, neighboring customers, and other non-participants. These distributional impacts are the primary focus of this chapter.

Indirect equity impacts of net metering refer to various second-order effects. If net metering encourages DG adoption, it might result in demand for particular types of labor and particular supply chain requirements compared to other energy resources. Distributed energy resources (DER) tend to generate and maintain more local jobs and more economic activity compared to traditional utility scale investments in energy generation and transmission. This is true for both higher numbers of jobs and higher economic multipliers per MW of capacity, per MWh of energy delivered or saved, and per million dollars of expenditure on each technology option.[19] In addition, the fact that 29.6 percent of the solar workforce is made up of women and 40 percent of the solar workforce are people of color[20] means that net metering policies hold the potential to shrink the racial employment gap relative to many other sectors of the energy workforce. However, the degree to which disadvantaged communities would share in these employment benefits without specific policy focus and program designs intended to ensure that result is not clear.[21] As a result, several states, including Illinois with its Clean Energy and Jobs Act (CEJA), require that local economic development, job training, and local job offerings be included in solar policies.[22]

At the same time, residents in disadvantaged communities are more vulnerable to losing their jobs if solar energy displaces fossil plant jobs in communities that have hosted

[19] EPA (U.S. Environmental Protection Agency). 2018. "Part Two: Chapter Five: Estimating the Economic Benefits of Energy Efficiency and Renewable Energy." In *Quantifying the Multiple Benefits of Energy Efficiency and Renewable Energy—A Guide for State and Local Governments*. https://www.epa.gov/statelocalenergy/quantifying-multiple-benefits-energy-efficiency-and-renewable-energy-guide-state.

[20] IREC (Interstate Renewable Energy Council). 2022. "National Solar Jobs Census 2021." https://irecusa.org/resources/national-solar-jobs-census-2021.

[21] See p. 134 in Leon, W., C. Farley, N. Hausman, B. Herbert, N.H. Hammer, B. Paulos, T. Reames, R. Sanders, L. Schieb, D. Deane-Ryan, and R. Navarra. 2019. *Solar with Justice: Strategies for Powering up Under-Resourced Communities and Growing an Inclusive Solar Market*. Montpelier, VT: Clean Energy States Alliance.

[22] More information about the Climate and Equitable Jobs Act, Public Act 102-0662 is available at: https://www2.illinois.gov/epa/topics/ceja/Pages/default.aspx.

coal plants.[23] Thus, there is evidence of additional distributional equity workforce issues, although they may not affect all regions equally.[24,25]

Societal equity effects of net metering refer to the distribution of externalities, which are the unpriced benefits and costs associated with BTM DG supported by net metering. For example, by promoting renewable or clean DG, net metering generally leads to the displacement of fossil fuel generation and reduces air pollutant emissions (see discussion in Chapter 4). These clean air benefits advantage society but will vary by region, neighborhood, and individual, resulting in varying equity outcomes. Since disadvantaged communities tend to suffer more from air pollution, clean energy policies that reduce pollution like net metering have potential to reduce pollution in overburdened communities, but the level of benefit will depend on factors such as the location of solar adoption, the location of emission reductions, and how these benefits compare to other clean energy adoption policies.

Dimensions of Equity

Equity may be characterized along four dimensions (among others): distributional, procedural, intergenerational, and structural. The following describes these dimensions in the broad context of energy and the specific context of net metering.

Distributional equity is the dimension of equity that has received the greatest attention. It refers to fairness in the allocation of benefits, costs, rights, resources, and information. Energy programs and policies achieve distributional equity when they result in fair distributions of benefits and costs across all segments of a community (including participants and non-participants), prioritizing those with the greatest need. Based on

---

[23] Carley, S., and D.M. Konisky. 2020. "The Justice and Equity Implications of the Clean Energy Transition." *Nature Energy* 5:569–577; Ross, E., M. Day, C. Ivanova, A. McLeod, and J. Lockshin. 2022. "Intersections of Disadvantaged Communities and Renewable Energy Potential: Data Set and Analysis to Inform Equitable Investment Prioritization in the United States." *Renewable Energy Focus* 41:1–14.

[24] The lack of a diverse solar workforce is prompting innovative programs. The Just Opportunity Circle, an organized group of U.S. economic development leaders, is exploring strategic solutions for equitable economic opportunities with a focus on low-wealth and communities of color. For more information, see PSE (Partnership for Southern Equity). 2021. "6 Principles of Economic Inclusion." https://psequity.org/just-opportunity. Solutions include policy changes that remove barriers to employment and hiring and training strategies that emphasize employment equity. For more information, see Louie, E.P., and J.M. Pearce. 2016. "Retraining Investment for U.S. Transition from Coal to Solar Photovoltaic Employment." *Energy Economics* 57:295–302.

[25] Brown, M.A., and M. Ahmadi. 2019. "Would a Green New Deal Add or Kill Jobs?" *Scientific American*. https://www.scientificamerican.com/article/would-a-green-new-deal-add-or-kill-jobs1.

different aspects of the energy grid, distributional equity can be looked at from two lenses: accessibility and affordability.

Distributional equity requires assessing which customers experience the benefits and costs of energy programs and investments. Traditional benefit-cost analysis (BCA) used to evaluate policies and programs has not typically examined such issues. Rather, it has focused on aggregate or average benefits and costs and has overlooked the distribution of impacts.

Understanding the impact of net metering on low-income households and disadvantaged communities requires distributional equity analysis (DEA), sometimes also referred to as distribution analysis or an analysis of distributional impacts. DEA is explicitly designed to measure the differential impacts between target populations and other customers. Analyses of rates, bills, participation, and other aspects of energy equity are core to DEA.[26,27] They can provide information on how impacts are experienced by host customers compared to other customers, which can reveal equity issues.[28,29] Such analyses, however, rely on adequate data collection and data transparency, which is unavailable in many regions. Table 5-1 provides a comparison of BCA and DEA as applied to net metering.

Ideally, analysis of an energy policy or program such as net metering would consider its standalone impact on society and its distributional impact on resource-constrained, marginalized, and energy cost-burdened customers, as well as disadvantaged and environmentally overburdened communities. Then net metering's impacts could be compared with the impact of other possible policies to support customer adoption of clean DG, considering their societal and distributional effects. This could enable policymakers and regulators to craft net metering variants and alternatives that, in addition to having a positive aggregate impact, may also have favorable distributional impacts. For example:

- If the principal policy goal is to minimize rate increases and household bills while achieving carbon reduction and other environmental goals, how does net metering fare with respect to distributional equity? Also, how does it compare to

---

[26] Pacific Northwest National Laboratory. "Energy Equity; PNNL's Vision Statement for Equity in the Power Grid." https://www.pnnl.gov/projects/energy-equity.

[27] University of Michigan School for Environment and Sustainability. 2022. "Energy Equity Project Report." https://energyequityproject.com/wp-content/uploads/2022/08/220174_EEP_Report_8302022.pdf.

[28] They can also be expanded to include information about target populations and other energy equity metrics such as Gini coefficients and Lorenz curves. See Sitthiyot, T., and K. Holasut, K. 2020. "A Simple Method for Measuring Inequality." *Palgrave Communications* 6(112):1–9.

[29] Johnson, E., R. Beppler, C. Blackburn, B. Staver, M. Brown, and D. Matisoff. 2017. "Peak Shifting and Cross-Class Subsidization: The Impacts of Solar PV on Changes in Electricity Costs." *Energy Policy* 106:436–444.

TABLE 5-1 Benefit-Cost Versus Distributional-Equity Analyses of Net Metering

	Benefit-Cost Analysis (BCA)	Distributional-Equity Analysis (DEA)
Purpose	To identify which DG utilities or policies should invest in or otherwise support on behalf of all customers	To identify how DG affects hosts versus non-host customers, and how it affects target populations as well as other customers
Questions Answered	What are the costs and benefits of DER across customers and perhaps society on average? How do these compare to other policies and resources? What are the costs and benefits of a DG program designed for target populations?	What is the impact of DG on hosts versus non-host customers, and on target populations as well as other customers?
Example Metrics for Reporting Results	Costs, Benefits, Net Benefits, and Benefit-Cost Ratios, using present-value dollars (PV$) Average impacts on rates ($/kWh), bills, and participation levels (% of eligible customers)	Impacts on target populations: Rate impacts ($/kWh) Bill impacts ($/month) Participation rates (% of eligible) Additional impacts: Energy burden Reliability Resilience Public health Job creation Community revitalization

SOURCE: Modified from NESP (National Energy Screening Project), 2022a, "Energy Equity and BCA," https://www.nationalenergyscreeningproject.org/resources/energy-equity-and-bca.

other clean energy policies in terms of their distributional effects? How can traditional net metering be modified to improve distributional equity?
- If the principal policy goal is to provide clean backup generation to vulnerable or disadvantaged populations, how does net metering compare to other policies in terms of their distributional impacts? Also, how does net metering compare to other programs that deliver equitable outcomes and services to disadvantaged communities, with respect to efficiency and cost- and rate-containment goals? In the communities that are served, does net metering or its policy alternatives further overburden non-participants through cost increases? How can it be revised to improve distributional equity?

Many additional concepts have been used to define and measure distributional equity aspects of the electricity grid, particularly the financial aspects of equity, important to the discussion of the equity of net metering policies. In the earlier stages of the electricity industry, equity was mainly measured in terms of accessibility to the grid. Now, considerations of distributional energy equity have been expanded and include the following concepts related to the affordability and cost of electricity:

- *Energy cost burden* (generally referred to as "energy burden") refers to the share of a household's income that is spent on energy utilities.[30]
- *Energy insecurity* is the uncertainty that a household might face in being able to make utility payments and the risk of shutoffs due to an inability to pay bills.
- *Energy poverty* means living in a home that does not have access to enough energy to meet essential needs.[31]

---

[30] Energy cost burden is typically measured as the ratio of the average household energy bill in a neighborhood, census tract, or customer class divided by the associated average household income. Energy cost burdens over 6 percent are typically considered unaffordably high. See Colton, R.D. 2011. "Home Energy Affordability in New York: The Affordability Gap (2008–2010)." Prepared for New York State Energy Research Development Authority. https://www.nyserda.ny.gov/-/media/Project/Nyserda/Files/EDPPP/LIFE/Resources/2008-2010-affordability-gap.pdf.

[31] Ibid. Colton (2011) defines "home energy burden" as those spending more than 6 percent of their income on meeting energy costs. The premise for this benchmark is that a household should not spend more than 30 percent of its income on housing expenses, and the utility costs should not exceed 20 percent of these expenses. This threshold is often used for comparison purposes and to estimate the "affordability gap." See Fisher, Sheehan & Colton. 2013. "The 2012 Home Energy Affordability Gap." www.homeenergyaffordabilitygap.com. A range of thresholds has also been developed. In a study for the State of Colorado, Cook and Shah (2018) distinguished between "energy stressed" households with energy burdens of 4–7 percent, "energy burdened" households with 7–10 percent energy burdens, and "energy impoverished" households with energy burdens greater than 10 percent. See Cook, J.J., and M. Shah. 2018. *Reducing Energy Burden with Solar: Colorado's Strategy and Roadmap for States*. NREL/TP-6A20-70965. Golden, CO: National Renewable Energy Laboratory.

Evaluating net metering policies requires comparing traditional net metering, its variants and policy alternatives in terms of both efficiency and equity. However, there is no consensus on how to design efficient and equitable rates, and net metering policies, in the context of increasing distributed generation.[32] The complexity of designing for both efficiency and equity is compounded by the four dimensions of equity and by the need to consider other factors such as environmental externalities, jobs, health impacts, customer decisions to disconnect from the grid, and resilience.

Procedural equity is fairness and transparency of the processes that allocate resources and resolve disputes. Neither BCA nor DEA reflect procedural equity.[33] However, both BCA and DEA can be enhanced by using fair and transparent processes to broadly engage constituents. Procedural equity focuses on who is represented, and who is engaged. An energy program or policy is procedurally equitable if it offers inclusive, accessible information (including information for customers and their advocates), as well as authentic engagement with and representation of all types of stakeholders and communities during the program's development, implementation, and evaluation, particularly for those who have not traditionally participated in energy policy and ratemaking processes. Energy justice[34] frameworks can help policymakers and regulators who set rates to make more informed decisions by following the principles of procedural justice.[35] This means identifying the priorities of the constituents who are being represented, how to balance economic efficiency with equity concerns, how DG investment costs should be covered, and how the benefits should be distributed. Failures of procedural justice can result in discrimination and marginalization, while failures of economic efficiency can result in higher rates. Procedural equity also embraces the processes by which rates may evolve over time, also called "transitional equity."[36] This report discusses procedural equity to the extent that it interacts with net metering policies.

[32] Rábago, K., and Valova, R. 2018. "Revisiting Bonbright's Principles of Public Utility Rates in a DER World." *The Electricity Journal* 31(8):9–13.

[33] NESP. 2022a. "Energy Equity and BCA." https://www.nationalenergyscreeningproject.org/resources/energy-equity-and-bca.

[34] Energy justice refers to the goal of achieving equity in both the social and economic participation in the energy system, while also remediating social, economic, and health burdens on those disproportionately harmed by the energy system. Energy justice explicitly centers the concerns of communities at the frontline of pollution and climate change ("frontline communities"), working class people, indigenous communities, and those historically disenfranchised by racial and social inequity. Energy justice aims to make energy accessible, affordable, clean, and democratically managed for all communities. For more information, see https://iejusa.org.

[35] Sovacool, B.K., and M.H. Dworkin. 2015. "Energy Justice: Conceptual Insights and Practical Applications." *Applied Energy* 142:435–444.

[36] Burger, S., I. Schneider, A. Botterud, and I. Pérez-Arriaga. 2018. *Fair, Equitable, and Efficient Tariffs in the Presence of Distributed Energy Resources.* Cambridge, MA: MIT Center for Energy and Environmental Policy Research.

Intergenerational equity adds a time dimension to the equity discussion by considering society's obligations to future generations. Discount rates are the analytic tool often used to enable the comparison of future and present benefits and costs,[37] a mechanism which can fold intergenerational equity into assessments of efficiency. Regulatory proceedings such as integrated resource planning quantify the value of benefits and costs to future generations when they specify discount rates to apply to their valuation. For example, applying low discount rates to assessments of clean energy benefits (i.e., discounting future benefits less) gives greater weight to the well-being of future generations. Actions that serve to increase rather than limit the options of future generations can be said to improve intergenerational equity.[38]

For example, energy policies and programs that reduce carbon emissions and, as a result, help mitigate climate change, contribute positively to intergenerational equity as they can reduce future costs due to climate destabilization. It should be noted, however, that lower-income households, particularly those with outstanding needs such as unpaid utility bills, may value current benefits more highly than possible future benefits, which would suggest a higher discount rate or a program focus on affordability. In some cases, current benefits can also reap future benefits. For instance, affordable energy via net-metered solar systems would reduce costs for these customers and allow low-income households to invest more now in human capital development through education and training, which could improve their earning potential in the future and across generations.

Structural equity involves making decisions with the recognition of historical, cultural, and institutional dynamics including structural racism, gender bias, and other dimensions of long-standing discrimination. For instance, the health and heat island consequences of highway development and deforestation in cities have caused capital to flow to more affluent neighborhoods. Excess heat and humidity drive the need for cooling and make the population more dependent on electricity and thus more vulnerable to outages. In such areas, DG in the form of microgrids or solar + storage-powered community resilience hubs and cooling centers may be particularly valuable.[39] Net metering compensation for this DG may or may not adequately reflect its increased value for such under-resourced communities.

[37] See Discount Rates section in Chapter 4.

[38] Norton, B.G. 2005. *Sustainability: A Philosophy of Adaptive Ecosystem Management*. Chicago, IL: University of Chicago Press.

[39] "Community resilience hubs" provide support for communities year-round and are equipped with solar + storage to provide backup power in the face of grid outages, providing vulnerable communities with access to cool air in hot weather, cell phone charging, refrigeration for medicines, power for medical equipment, and other critical services. For more information, see https://www.usdn.org/resilience-hubs.htm.

Redlining and discriminatory bank lending are also examples of structural equity issues that have limited BTM DG adoption by Black households in American cities, due to lack of access to capital, which has led to lower home ownership and a lack of home repair.[40] Lower levels of home ownership have blocked many communities of color from building wealth, presenting barriers to financing BTM solar and other DER because of their high up-front costs, which traditional net metering cannot overcome.[41] Compounding this effect, renters also face barriers due to limited control over decisions to install DG, limiting their ability to participate in net metering.

THE DEMOGRAPHICS OF NET METERING PARTICIPATION

Net metering can play a role in alleviating—or exacerbating—existing inequities associated with power supply and delivery across the United States. This section begins by describing the systemic inequities that have resulted in lower adoption rates of rooftop solar in low- and moderate-income communities, populations of color, and among renters.[42,43,44] The discussion also addresses adoption rates in communities with high housing burden, linguistic isolation, higher inequality, and lower educational attainment.[45] Net metering can interact with these systemic inequities to affect these populations, including impacts on income, affordability, and the ability to adopt BTM DG and participate in other programs. Attention then turns to access to solar and current net metering participation levels for renters, populations of color, and other subgroups, and barriers to participation for these communities. Finally, the discussion addresses strategies to alleviate these impacts and reduce historic barriers to BTM DG adoption through

[40] Sunter, D.A., S. Castellanos, and D.M. Kammen. 2019. "Disparities in Rooftop Photovoltaics Deployment in the United States by Race and Ethnicity." *Nature Sustainability* 2(1):71–76.

[41] Wilson, B. 2020. "Urban Heat Management and the Legacy of Redlining." *Journal of the American Planning Association* 86(4):443–457.

[42] Sunter, D.A., S. Castellanos, and D.M. Kammen. 2019. "Disparities in Rooftop Photovoltaics Deployment in the United States by Race and Ethnicity." *Nature Sustainability* 2(1):71–76.

[43] Forrester, S., G. Barbose, E. O'Shaughnessy, N. Darghouth, and C.C. Montañés. 2022. *Residential Solar-Adopter Income and Demographic Trends: November 2022 Update*. Berkeley, CA: Lawrence Berkeley National Laboratory.

[44] Gao, X., and S. Zhou. 2022. "Solar Adoption Inequality in the U.S.: Trend, Magnitude, and Solar Justice Policies." *Energy Policy* 169(October):113–163.

[45] Darghouth, N.R., E. O'Shaughnessy, S. Forrester, and G. Barbose. 2022. "Characterizing Local Rooftop Solar Adoption Inequity in the US." *Environmental Research Letters* 17(3):034028; Lukanov, B.R., and E.M. Krieger. 2019. "Distributed Solar and Environmental Justice: Exploring the Demographic and Socio-Economic Trends of Residential PV Adoption in California." *Energy Policy* 134:110935; Yu, J., Z. Wang, A. Majumdar, and R. Rajagopar. 2018. "DeepSolar: A Machine Learning Framework to Efficiently Construct a Solar Deployment Database in the United States." *Joule* 2(12):2605–2617.

net metering policies. This context is critical for understanding how net metering may improve or exacerbate existing inequities.

Low-Income Households

Low-income households face significantly higher energy cost burdens than the population at large,[46] with low-income U.S. households measured as spending an average of 8.1 percent of their income on energy costs, compared with an average of 2.3 percent for those not considered low-income.[47] In some areas low-income households that use less electricity than average also pay higher de facto rates for electricity under rate designs with high fixed monthly charges. Low-income households may struggle to afford utility bills among other expenses and may therefore face pressure to reduce their energy use to thermally uncomfortable levels.[48] Electricity shutoffs that result from utility debt can exacerbate both physical and mental health,[49] and low incomes are linked to a higher likelihood of mortality during extreme heat events.[50]

Lower incomes are also linked to living in inefficient buildings that are expensive to heat and cool.[51] Barriers to clean energy adoption include lack of quality internet access, which lags in low income and rural areas[52] and is sometimes necessary to implement

[46] Kontokosta, C.E., V.J. Reina, and B. Bonczak. 2020. "Energy Cost Burdens for Low-Income and Minority Households." *Journal of the American Planning Association* 86(1):89–105.

[47] American Council for an Energy-Efficient Economy. 2020. "Report: Low-Income Households, Communities of Color Face High 'Energy Burden' Entering Recession." https://www.aceee.org/press-release/2020/09/report-low-income-households-communities-color-face-high-energy-burden.

[48] Anderson, W., V. White, and A. Finney. 2012. "Coping with Low Incomes and Cold Homes." *Energy Policy* 49:40–52; Hernández, D. 2016. "Understanding 'Energy Insecurity' and Why It Matters to Health." *Social Science & Medicine* 167:1–10.

[49] Hernández, D. 2016. "Understanding 'Energy Insecurity' and Why It Matters to Health." *Social Science & Medicine* 167:1–10.

[50] Harlan, S.L., J.H. Declet-Barreto, W.L. Stefanov, and D.B. Petitti. 2013. "Neighborhood Effects on Heat Deaths: Social and Environmental Predictors of Vulnerability in Maricopa County, Arizona." *Environmental Health Perspective* 121(2):197–204; Sakka, A., M. Santamouris, I. Livada, F. Nicol, and M. Wilson. 2012. "On the Thermal Performance of Low Income Housing During Heat Waves." *Energy Build* 49:69–77.

[51] Brown, M.A., A. Soni, M.V. Lapsa, and K.A. Southworth. 2020a. *Low-Income Energy Affordability: Conclusions from a Literature Review.* Oak Ridge, TN: Oak Ridge National Laboratory; Gillard, R., C. Snell, and M. Bevan. 2017. "Advancing an Energy Justice Perspective of Fuel Poverty: Household Vulnerability and Domestic Retrofit Policy in the United Kingdom." *Energy Research & Social Science* 29:53–61; Hernández, D. 2016. "Understanding 'Energy Insecurity' and Why It Matters to Health." *Social Science & Medicine* 167:1–10; Walker, G., and R. Day. 2016. "Necessary Energy Uses and a Minimum Standard of Living in the United Kingdom: Energy Justice or Escalating Expectations?" *Energy Research & Social Science* 18:129–138.

[52] Martin, M. 2021. *For the First Time, Census Bureau Data Show Impact of Geography, Income on Broadband Internet Access.* Washington, DC: U.S. Census Bureau. https://www.census.gov/library/stories/2018/12/rural-and-lower-income-counties-lag-nation-internet-subscription.html.

rooftop solar. Income-constrained households also have limited access to smart appliances and thermostats to adjust their consumption, which may make solar and net metering less valuable to them. As mentioned, lack of home ownership and access to financing constitute other barriers, though it should be noted that these are not always multifamily buildings or rented, as approximately 54 percent of low-income households live in single-family homes, 69 percent of which are owner-occupied.[53] Those facing barriers to adopting DG inherently face barriers to participating in net metering and are thus less likely to benefit from it. Customers receiving service on discounted low-income utility rates may also be compensated under net metering tariffs at those discounted rates—that is, solar supplied back to the grid is compensated at discounted rather than retail rates—resulting in longer payback times for investments in DG.

For participants, BTM DG supported by net metering holds the potential both to reduce energy costs[54] and provide bill stability, which may have value for populations facing high energy cost burdens and energy insecurity. However, many barriers hinder the adoption of DG and exacerbate inequities for under-resourced groups.[55] As a result, programs aimed at boosting DG adoption in disadvantaged communities need to consider structural barriers beyond income to address race, language, education, historic investments, home ownership, access to capital and credit, and other factors.[56,57]

---

[53] DOE (Department of Energy). n.d. "Clean Energy for Low Income Communities: Single Family Home Retrofits." https://betterbuildingssolutioncenter.energy.gov/CELICA-Toolkit/single-family-home-retrofits.

[54] Darghouth, N.R., G. Barbose, J. Zuboy, P.J. Gagnon, A.D. Mills, and L. Bird. 2020. "Demand Charge Savings from Solar PV and Energy Storage." *Energy Policy* 146(November):111766.

[55] Brown, M.A., J. Hubbs, X.V. Gu, and M.-K. Cha. 2021. "Rooftop Solar for All: Closing the Gap Between the Technically Possible and the Achievable Potentials." *Energy Research and Social Science* 80(October):102203; Scavo, J., S. Korosec, E. Guerrero, B. Pennington, and P. Doughman. 2016. *Low-Income Barriers Study, Part A: Overcoming Barriers to Energy Efficiency and Renewables for Low-Income Customers and Small Business Contracting Opportunities in Disadvantaged Communities*. Publication Number: CEC-300-2016-009-CMF. Sacramento, CA: California Energy Commission; Sunter, D.A., S. Castellanos, and D.M. Kammen. 2019. "Disparities in Rooftop Photovoltaics Deployment in the United States by Race and Ethnicity." *Nature Sustainability* 2(1):71–76.

[56] Scavo, J., S. Korosec, E. Guerrero, B. Pennington, and P. Doughman. 2016. *Low-Income Barriers Study, Part A: Overcoming Barriers to Energy Efficiency and Renewables for Low-Income Customers and Small Business Contracting Opportunities in Disadvantaged Communities*. Publication Number: CEC-300-2016-009-CMF. Sacramento, CA: California Energy Commission.

[57] Darghouth et al. (2022) found higher adoption rates among LMI communities with LMI-focused installers and installers providing leases. See Darghouth, N.R., E. O'Shaughnessy, S. Forrester, and G. Barbose. 2022. "Characterizing Local Rooftop Solar Adoption Inequity in the US." *Environmental Research Letters* 17(3):034028. O'Shaughnessy et al. (2020) also found that LMI-focused incentives could help reduce solar adoption inequities, while standard incentives risked exacerbating these inequities. See O'Shaughnessy, E., G. Barbose, R. Wiser, S. Forrester, and N. Darghouth. 2020 "The Impact of Policies and Business Models on Income Equity in Rooftop Solar Adoption." *Nature Energy* 6(1):84–91.

Virtual net metering and community solar can help to provide access to solar for these customer groups. Some utilities have proposed and offer community solar subscription programs to address the issue of affordability. The subscription programs would allow low-income households to purchase community solar at a discounted price, with the difference between the reduced price and the retail rate being covered by donations from other ratepayers, community organizations and businesses (see community solar discussion below and in Box 3-1). Low-income customers could also potentially participate in community-level projects and campaigns led by rooftop-solar advocates,[58] such as Solarize, which is a community-based program that aims to make solar panels more affordable for community members through grassroots-level organizing and group purchasing.[59] While combining net metering with low- and moderate-income (LMI)-focused programs, such as community solar, may help reduce some of these inequities and improve access for historically underserved populations,[60] these programs are not likely to fully address distributional equity concerns related to the possibility of costs shifting from net metering participants to non-participants.

---

[58] Weissman, G. 2017. "Rental Rooftops: A Missed Opportunity for Solar Energy." https://frontiergroup.org/blogs/blog/fg/rental-rooftops-missed-opportunity-solar-energy.

[59] "Solarize facilitates solar-panel purchases by prescreening and selecting an installer to provide services to its residents, taking advantage of economies of scale from bulk purchasing. Each Solarize campaign typically involves: (1) group purchasing and a decreasing tiered pricing schedule for materials; (2) a competitive bidding process; (3) free evaluations with a financial analysis, preliminary site design, and energy-usage profile; (4) free workshops on the benefits of solar that allow participants to meet with contractors; and (5) a limited time window to encourage participants to enroll quickly and generate momentum for the campaign. To ensure that community members can afford rooftop installations, Solarize also partners with not-for-profit credit unions to provide loans specifically targeted toward clean energy. Community organizers may initiate a Solarize campaign by starting a petition among community members" (Brown et al. 2021, p. 3). Brown, M.A., J. Hubbs, X.V. Gu, and M.-K. Cha. 2021. "Rooftop Solar for All: Closing the Gap Between the Technically Possible and the Achievable Potentials." *Energy Research and Social Science* 80(October):102203. For more information, see Solarize campaigns, Solar CrowdSource at https://www.solarcrowdsource.com.

[60] See, for example, Paulos et al. 2021. "An Assessment of Evaluation Practices of Low- and Moderate-Income Solar Programs." Berkeley, CA: Lawrence Berkeley National Laboratory; Stanton, T. 2020. *Solar Energy That Pays for Low-Income Customers and Communities*. NRRI Insights Paper. Silver Spring, MD: National Regulatory Research Institute. https://pubs.naruc.org/pub/46965D7D-155D-0A36-315D-58319B591EB8; and Zinaman, O.R., T. Bowen, and A.Y. Aznar. 2020. *An Overview of Behind-the-Meter Solar-Plus-Storage Regulatory Design: Approaches and Case Studies to Inform International Applications*. Report No. NREL/TP-7A40-75283. Golden, CO: National Renewable Energy Laboratory.

Populations of Color

People of color face greater procedural injustices than other customers in the forms of (1) discrimination related to housing, employment, and credit; (2) lack of informed consent for energy projects; (3) lack of representation in decision-making processes; and (4) lack of access to information. People of color may also face greater health impacts tied to the inability to cool homes, and they have a higher likelihood of death from extreme heat events.[61,62,63]

Historic underinvestment in infrastructure in communities of color and other disadvantaged communities can also contribute to ongoing energy inequities. For example, the distribution grid in California has been found to have a lower hosting capacity[64] for DER on circuits with higher proportions of Black households and disadvantaged communities.[65] Similarly, satellite imagery of Texas in February 2021 found that widespread outages caused by winter storms were four times more prevalent in minority communities than in predominantly White communities.[66] Households of color frequently face higher energy cost burdens[67] and greater energy insecurity[68] than White households, not only because of their lower average incomes but also because of a history of discriminatory

[61] Pager, D., and H. Shepherd. 2008. "The Sociology of Discrimination: Racial Discrimination in Employment, Housing, Credit, and Consumer Markets." *Annual Review of Sociology* 34:181–209.

[62] Sovacool, B.K., and M.H. Dworkin. 2015. "Energy Justice: Conceptual Insights and Practical Applications." *Applied Energy* 142:435–444.

[63] Harlan, S.L., J.H. Declet-Barreto, W.L. Stefanov, and D.B. Petitti. 2013. "Neighborhood Effects on Heat Deaths: Social and Environmental Predictors of Vulnerability in Maricopa County, Arizona." *Environmental Health Perspective* 121(2):197–204.

[64] While there is no formal technical definition for *hosting capacity*, this important concept can be understood as the amount of rooftop solar that can be safely installed at the distribution system level before requiring system upgrades such as additional supervisory control and data acquisition (SCADA), increasing capacity of lines, or addition of advanced voltage regulating devices.

[65] Brockway, A.M., J. Conde, and D. Callaway. 2021. "Inequitable Access to Distributed Energy Resources Due to Grid Infrastructure Limits in California." *Nature Energy* 6:892–903.

[66] Carvallo, J.P., F.C. Hsu, Z. Shah, and J. Taneja. 2021. *Frozen Out in Texas: Blackouts and Inequity*. New York: The Rockefeller Foundation.

[67] Kontokosta, C.E., V.J. Reina, and B. Bonczak. 2020. "Energy Cost Burdens for Low-Income and Minority Households." *Journal of the American Planning Association* 86(1):89–105.

[68] Graff, M., S. Carley, D.M. Konisky, and T. Memmott. 2021. "Which Households Are Energy Insecure? An Empirical Analysis of Race, Housing Conditions, and Energy Burdens in the United States." *Energy Research & Social Science* 79:102144; Memmott, T., S. Carley, M. Graff, and D.M. Konisky. 2021. "Sociodemographic Disparities in Energy Insecurity Among Low-Income Households Before and During the COVID-19 Pandemic." *Nature Energy* 6:186–193.

lending practices that have resulted in a lack of access to capital. These conditions can limit their ability to purchase rooftop solar systems and benefit from net metering.[69]

Renters

Residents of multi-family and renter-occupied properties face additional barriers to BTM DG adoption and participation in net metering.[70] With both master metering and submetering,[71] the landlord is often responsible for appliances and other energy infrastructure within the complex but does not pay the energy bill. Renters are typically not able to add solar infrastructure to homes or buildings they do not own. These groups also may move more often, which makes conventional PV systems ownership impractical. Renters could install "plug-and-play" PV systems that are generally PV modules with microinverters.[72] Such "balcony solar" systems are widespread in Europe, and there is a potential market for them in the United States because of their ability to reduce renters' electricity costs.

The resulting "split incentive" problem may present a barrier to the adoption of clean energy (including DG supported by net metering) for renter households and contributes to their higher energy bills;[73] however, environmentally committed tenants may have the option to act themselves through investments in community and other solar projects where virtual net metering (VNM) is available. Renters also tend to be lower-income, are

[69] As documented by Sunter et al. (2019), which found significant racial and ethnic differences in rooftop solar adoption in the United States. See Sunter, D.A., S. Castellanos, and D.M. Kammen. 2019. "Disparities in Rooftop Photovoltaics Deployment in the United States by Race and Ethnicity." *Nature Sustainability* 2(1):71–76.

[70] Heeter, J., A. Sekar, E. Fekete, M. Shah, and J.J. Cook. 2021. *Affordable and Accessible Solar for All: Barriers, Solutions, and On-Site Adoption Potential.* Golden, CO: National Renewable Energy Laboratory. https://www.nrel.gov/docs/fy21osti/80532.pdf; Mateyka, P.J., and C.R. Mazur. 2021. *Homeownership in the United States: 2005 to 2019.* Washington, DC: U.S. Census Bureau. https://www.census.gov/content/dam/Census/library/publications/2021/acs/acsbr-010.pdf.

[71] "Master metering" measures the electricity use for an entire rental property (with utility costs typically charged to renters as a fixed dollar amount, part of their rent) and "submetering" measures the electricity consumption for each rental unit individually.

[72] Mundada, A.S., E.W. Prehoda, and J.M. Pearce. 2017. "U.S. Market for Solar Photovoltaic Plug-and-Play Systems." *Renewable Energy* 103:255–264.

[73] Melvin, J. 2018. "The Split Incentives Energy Efficiency Problem: Evidence of Underinvestment by Landlords." *Energy Policy* 115:342–352.

more likely to be people of color, less likely to have access to energy-efficient appliances, and more likely to have higher energy cost burdens.[74]

Vulnerable Populations

Vulnerable populations include the elderly, who require a narrow band of home temperatures;[75] individuals who rely on electricity to power medical devices; and households with young children. The energy requirements of these vulnerable populations typically result in higher energy bills and may cause energy insecurity.[76] Low-income families with young children experience heightened financial pressures associated with ensuring that their children are well fed and healthy,[77] and in addition, face greater competing childcare expenses, and hence more challenges to meet their energy needs.[78]

[74] Drehobl, A., L. Ross, and R. Ayala. 2020. *How High Are Household Energy Burdens? An Assessment of National and Metropolitan Energy Burden Across the United States*. Washington, DC: American Council for an Energy-Efficient Economy; Lang, M., R. Lane, K. Zhao, S. Tham, K. Woolfe, and R. Raven. 2021. "Systematic Review: Landlords' Willingness to Retrofit Energy Efficiency Improvements." *Journal of Cleaner Production* 303:127041; Lukanov, B., A. Makhijani, K. Shetty, Y. Kinkhabwala, A. Smith, and E. Krieger. 2022. *Pathways to Energy Affordability in Colorado*. Oakland, CA: PSE Healthy Energy, and Takoma Park, MD: Institute for Energy and Environmental Research. https://www.psehealthyenergy.org/wp-content/uploads/2022/01/Colorado-Energy-Affordability-Study_Full-Report.pdf.

[75] Basu, R. 2009. "High Ambient Temperature and Mortality: A Review of Epidemiologic Studies from 2001 to 2008." *Environmental Health* 8:40; Day, R., G. Walker, and N. Simcock. 2016. "Conceptualising Energy Use and Energy Poverty Using a Capabilities Framework." *Energy Policy* 93:255–264; Knowlton, K., M. Rotkin-Ellman, G. King, H.G. Margolis, D. Smith, G. Solomon, R. Trent, and P. English. 2009. "The 2006 California Heat Wave: Impacts on Hospitalizations and Emergency Department Visits." *Environmental Health Perspective* 117(1):61–67; O'Shaughnessy, E., G. Barbose, R. Wiser, S. Forrester, and N. Darghouth. 2020. "The Impact of Policies and Business Models on Income Equity in Rooftop Solar Adoption." *Nature Energy* 6(1):84–91; Ormandy, D., and Ezratty, V. 2012. "Health and Thermal Comfort: From WHO Guidance to Housing Strategies." *Energy Policy* 49:116–121; Vandentorren, S., P. Bretin, A. Zeghnoun, L. Mandereau-Bruno, A. Croisier, C. Cochet, J. Ribéron, I. Siberan, B. Declercq, and M. Ledrans. 2006. "August 2003 Heat Wave in France: Risk Factors for Death of Elderly People Living at Home." *European Journal of Public Health* 16:583–591.

[76] Day, R., G. Walker, and N. Simcock. 2016. "Conceptualising Energy Use and Energy Poverty Using a Capabilities Framework." *Energy Policy* 93:255–264; Memmott, T., S. Carley, M. Graff, and D.M. Konisky. 2021. "Sociodemographic Disparities in Energy Insecurity Among Low-Income Households Before and During the COVID-19 Pandemic." *Nature Energy* 6:186–193; Middlemiss, L., and R. Gillard. 2015. "Fuel Poverty from the Bottom-Up: Characterising Household Energy Vulnerability Through the Lived Experience of the Fuel Poor." *Research & Social Science* 6:146–154.

[77] Hernández, D. 2016. "Understanding 'Energy Insecurity' and Why It Matters to Health." *Social Science & Medicine* 167:1–10; Nord, M., and L.S. Kantor. 2006. "Seasonal Variation in Food Insecurity Is Associated with Heating and Cooling Costs Among Low-Income Elderly Americans." *Journal of Nutrition* 136:2939–2944.

[78] Cook, J.T., P.H. Casey, R. Rose-Jacobs, M.M. Black, M. Chilton, S. Ettinger de Cuba, D. Appugliese, S. Coleman, T. Heeren, C. Berkowitz, and D.B. Cutts. 2008. "A Brief Indicator of Household Energy Security: Associations with Food Security, Child Health, and Child Development in US Infants and Toddlers." *Pediatrics* 122(4):e867–e875.

Young children are also at higher risk of morbidity in extreme weather events.[79] After controlling for income, Black households with children were more likely than any other group to experience energy insecurity.[80]

Net Metering, Rate Structures, and Demand Flexibility

Time-of-use (TOU) rates, which reflect the varying costs of electricity generation at different times of day, can be a rate design that better aligns prices with underlying costs (see Chapter 4). They have been applied to net metering customers (e.g., in California, net metering 2.0 used to requires all customers to adopt TOU rates[81]) to encourage the use of electricity when low-cost renewable sources are readily available (e.g., solar in the mid-afternoon as the peak period for PG&E is 4–9 pm on weekdays after solar production drops).[82]

The literature indicates mixed results about the impact of TOU rates on low- and moderate-income (LMI) customers (e.g., because they cannot choose their appliances or afford smart ones or otherwise meaningfully modify their electricity usage patterns).[83] Some analyses show that LMI customers who have opted in to TOU rates respond similarly to non-LMI customers, and are as satisfied or more satisfied with their savings (on

[79] Basu, R. 2009. "High Ambient Temperature and Mortality: A Review of Epidemiologic Studies from 2001 to 2008." *Environmental Health* 8:40.

[80] Hernández, D., Y. Aratani, and Y. Jiang. 2014. *Energy Insecurity Among Families with Children*. New York: National Center for Children in Poverty, Columbia University.

[81] An issue unique to California is that qualifying lower income customers were paid less for exporting rooftop solar electricity than other residential NEM 2.0 participants. This is because qualifying lower income customers can take service on a 30–35 percent discounted electric rate, called CARE. As NEM 2.0 compensated customers based on the retail rate, it also compensated CARE customers 30–35 percent less than non-CARE customers. NEM 3.0 adjusts for this by providing an export adder for CARE customers; to further encourage installation of distributed generation in disadvantaged communities, the NEM 3.0 Decision recommended that approximately $600 million dollars from the state budget be used to subsidize rooftop solar and storage in disadvantaged communities. Time will tell whether these policy updates can adequately encourage adoption in disadvantaged communities in California.

[82] For additional information on net metering 2.0 in California, see https://www.cpuc.ca.gov/industries-and-topics/electrical-energy/demand-side-management/net-energy-metering. In states with limited rooftop solar, the peak period with highest prices may be in the afternoon (e.g., 2–7 pm for Georgia on summer weekdays). For information about time-of-use rates for PG&E and Georgia Power, see https://www.pge.com/en_US/residential/rate-plans/rate-plan-options/time-of-use-base-plan/tou-b.page and https://www.georgiapower.com/residential/billing-and-rate-plans/pricing-and-rate-plans/nights-weekends.html.

[83] White, L.V., and N.D. Sintov. 2020. "Health and Financial Impacts of Demand-Side Response Measures Differ across Sociodemographic Groups." *Nature Energy* 5:50–60.

a percentage basis).[84] This suggests that there are ways to design TOU rates to avoid adverse impacts on vulnerable populations. One example may be shadow billing, which has been used for analysis of TOU rate impacts and could be considered to hold customers harmless from increases.[85] Customers without access to the internet may face a barrier to adopting DG and qualifying for net metering, which may also hinder their ability to take advantage of or benefit from TOU rate design. More study of the effect of TOU rates on low- and moderate-income customers and their ability to adopt BTM DG is needed.

EQUITY CONCERNS OF NET METERING NON-PARTICIPATION

Net metering participants may be over- or under-compensated for the value their BTM DG provides to the system, depending on factors such as the level of a utility's retail rates and the magnitude of solar penetration in a utility's service territory, as discussed in Chapter 4. This can have cost consequences for net metering non-participants.

- On the one hand, net metering participants may be over-compensated, and non-participants may see upward pressure on their rates and costs when (1) net metering rates are equal to retail prices, (2) retail prices exceed the value of net metering customers' exports to the system, and (3) net metering participant payments to the utility do not cover the utility's costs to provide service to them. In such circumstances, costs may shift to non-participating customers whose energy bills could rise. Net metering non-participants could thereby experience an additional cost burden that may or may not be greater than the value of the societal benefits enabled by net metering participants.
- On the other hand, net metering participants may be under-compensated and non-participants may see downward pressure on their rates and costs when (1) rates and levels of solar penetration are low, and (2) solar is coincident with system demand peaks. In such circumstances, solar adoption by net metering participants may help defer transmission, distribution, and capacity investments, thereby reducing these costs that would otherwise be included in rates. Depending on the source of grid power, net metering participants can even be heavily subsidizing non-participants because they are reducing the cost of pollution and high-cost operation of coal plants. In addition, customers who were encouraged

[84] Faruqui, A., N. Powers, and S. Sergici. 2020. *Study by Brattle Economists Evaluates Time-of-Use (TOU) Pilots for Maryland Utilities*. Prepared for Maryland Joint Utilities. Brattle Group. https://www.brattle.com/insights-events/publications/study-by-brattle-economists-evaluates-time-of-use-tou-pilots-for-maryland-utilities.

[85] Migden-Ostrander, J. 2017. "Utility 'Shadow Billing' Can Shed Light on Rate Options." *RAP*, August 29. https://www.raponline.org/blog/utility-shadow-billing-can-shed-light-on-rate-options.

by net metering to invest in BTM DG, such as solar, may help drive down cost curves for these technologies and enable a broader customer base to afford them.[86]

Methods to achieve an equitable outcome may vary across electricity service territories, depending on rate design and the level of net metering participation, as well as attributes of the utility and its customers' loads. Keeping this complex set of factors in mind, an equitable rate design could ensure that net metering customers with BTM resources pay the net costs to serve them (accounting for system benefits), and non-participants do not have to make up for any shortfalls. Because net metering customers tend to have higher incomes than other customers,[87] if costs are shifted to non-participants, on average lower-income households would be subsidizing wealthier customers. The equity implications of rate design, inclusive of any cost shifts, might be considered for all rate components, not just those that may be associated with net metering.

Net Metering Participation and Equity Impacts on Electricity Bills and Rates

Any increases in rates and bills are particularly harmful to low-income communities and energy-burdened customers.[88] Increasing BTM DG adoption, which reduces consumption of utility-provided electricity, can put upward pressure on retail electricity rates and bills, particularly where costs are largely recovered in volumetric charges that differ from social marginal costs (see Chapter 4).[89] The impact on rates and bills (and hence on equity) depends on the penetration of rooftop solar in the utility's service territory.[90] For most states, the penetration of BTM DG and therefore the impact of net metering

[86] Speelman, L., and Y. Numata. 2022. "Harnessing the Power of S-Curves: How S-Curves Work and What We Can Do to Accelerate Them." https://rmi.org/insight/harnessing-the-power-of-s-curves.

[87] Forrester, S., G. Barbose, E. O'Shaughnessy, N. Darghouth, and C.C. Montañés. 2022. *Residential Solar-Adopter Income and Demographic Trends: November 2022 Update*. Berkeley, CA: Lawrence Berkeley National Laboratory.

[88] Borenstein, S., M. Fowlie, and J. Sallee. 2021. *Designing Electricity Rates for an Equitable Energy Transition*. Berkeley, CA: Energy Institute at HAAS. https://haas.berkeley.edu/wp-content/uploads/WP314.pdf; Penn, I. 2022. "A Fight Over Rooftop Solar Threatens California's Climate Goals." *The New York Times*, January 24. https://www.nytimes.com/2022/01/24/business/energy-environment/california-rooftop-solar-utilities.html.

[89] Murray, B. 2019. "The Paradox of Declining Renewable Costs and Rising Electricity Prices." *Forbes*, June 17, https://www.forbes.com/sites/brianmurray1/2019/06/17/the-paradox-of-declining-renewable-costs-and-rising-electricity-prices.

[90] Barbose, G., and N. Darghouth. 2019. *Tracking the Sun: Pricing and Design Trends for Distributed Photovoltaic Systems in the United States*. Berkeley, CA: Lawrence Berkeley National Laboratory. https://emp.lbl.gov/sites/default/files/tracking_the_sun_2019_report.pdf.

on average retail electricity prices is small.[91] In states or utilities with high distributed solar penetration levels and traditional net metering, the impacts on non-participants' electricity rates and bills may be significant but depend critically on underlying rate structures.[92] At low participation levels (as currently exist and are projected to persist in the vast majority of places in the United States), even with a substantial mismatch between net metering compensation rates, the rate impacts are likely to be small.[93] See Figure 5-2.

Disconnection from the Grid

As discussed in Chapter 4, changes to net metering policy could inadvertently lead to customer decisions to disconnect or defect from the grid for those who can afford it. Equity issues can arise when customers disconnect from the grid, resulting in the remaining customers paying the full cost of grid maintenance and operation, which could cause

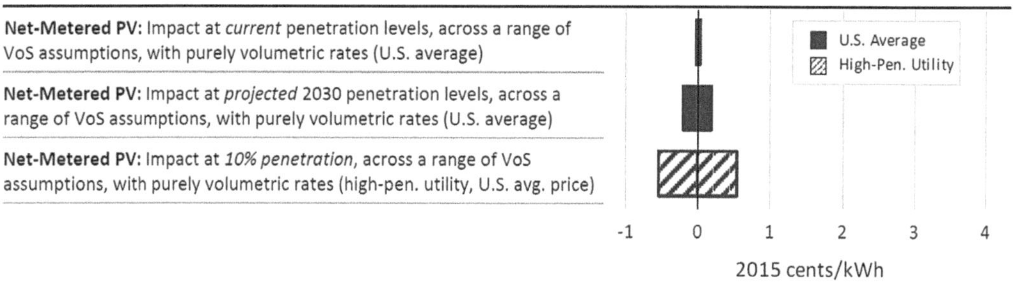

FIGURE 5-2 Effects of net metering on average retail electricity prices.
SOURCE: Adapted from G. Barbose, 2017, p. 28 in *Putting the Potential Rate Impacts of Distributed Solar into Context*, Berkeley, CA: Lawrence Berkeley National Laboratory, https://emp.lbl.gov/publications/putting-potential-rate-impacts.

---

[91] See Chapter 3 for more information.

[92] See p. 67 in NARUC (National Association of Regulatory Utility Commissioners). 2016. *Distributed Energy Resources Rate Design and Compensation*. Washington, DC: NARUC. https://pubs.naruc.org/pub.cfm?id=19FDF48B-AA57-5160-DBA1-BE2E9C2F7EA0. See also Barbose, G. 2017. *Putting the Potential Rate Impacts of Distributed Solar into Context*. Berkeley, CA: Lawrence Berkeley National Laboratory. https://emp.lbl.gov/publications/putting-potential-rate-impacts and Lawson, A.J. 2019. *Net Metering: In Brief*. CRS Report R46010. Washington, DC: Congressional Research Service.

[93] Barbose, G. 2017. *Putting the Potential Rate Impacts of Distributed Solar into Context*. Berkeley, CA: Lawrence Berkeley National Laboratory. https://emp.lbl.gov/publications/putting-potential-rate-impacts.

their effective rates to increase. This is of particular concern because these remaining customers are likely to be disproportionately low income.[94,95]

ENVIRONMENT, HEALTH, AND OTHER SOCIETAL EQUITY EFFECTS OF NET METERING

The total capacity, technology type, and spatial distribution of BTM DG adoption hold multiple implications for the environment, public health, and equity. On a system level, solar and other BTM DG can displace fossil fuel generation and reduce associated greenhouse gas (GHG) and health-damaging air pollutant emissions, although these benefits may impact some regions and neighborhoods more than others. On an individual household level, an increase in household energy security and resilience for rooftop solar adopters can also have public health benefits. As discussed in Chapter 4, net metering tariffs can be designed to reflect some of these benefits, but these are often evaluated in aggregate. Here, we also look at equity dimensions related to the *distribution* of these benefits.

System Level Air Pollutant Emissions Impacts on Equity

As described in Chapter 4, fossil fuel combustion releases both GHGs and health-damaging air pollutants (fine particulates, sulfur dioxide, and nitrogen oxides). While the impact of GHGs is global (the location of the source is irrelevant), the public health impacts of air pollutants depend on when and where they are emitted. Where DG is adopted will also affect the location of the marginal power plant emissions displaced, and therefore where public health benefits may be realized. This holds implications for how externalities may be incorporated into the value of solar calculations and net metering compensation rates. For example, siting rooftop solar within a transmission-constrained load pocket,

[94] The committee observes that some people sometimes draw analogies between disconnection from the grid and the implications of community choice aggregation (CCA), even though joining a CCA and participating in net metering are not related. In states that allow it (e.g., Massachusetts and California), CCAs may form in communities and provide electricity supply to the members of that community even though the supply is still delivered by the local distribution utility. Joining a CCA can give rise to situations where the participating customers do not pay all the costs the utility incurred on their behalf prior to their joining the CCA. Such situations can lead to cost shifts from participating customers to the customers that remain with the utility for bundled service even though they have nothing to do with net metering tariffs and any cost shifts related to them.

[95] Castanda, M., M. Jimenez, S. Zapata, C.J. Franco, and I. Dyner. 2017. "Myths and Facts of the Utility Death Spiral." *Energy Policy* 110:105–116.

found in many urban areas, is more likely to displace demand for fossil fuel generation, and associated air pollutants, located within that load pocket (e.g., inefficient peaker plants with high emissions rates that meet local demand when electricity cannot be delivered from the grid at large). While solar that offsets marginal generation from a power plant far away may have GHG reduction benefits, it would not reduce local air pollution.

Another consideration is the impact of DG on the operations of and emissions from specific power plants. While distributed solar that displaces fossil generation reduces GHG and air pollutant emissions overall, widespread adoption of variable technologies, such as solar, may increase the need in certain locations for rapid ramping of power plants when the sun sets or is covered by clouds, causing an increase in emissions than would otherwise occur with steady-state operation.[96,97,98] The distribution of public health impacts of power plant emissions (and the benefit of a reduction in emissions) therefore depends on the location of the power plant, other generation on that system, the magnitude of emitted pollutants, and the characteristics of the populations living near and downwind from the plant. Vulnerable populations such as children, the elderly, those with underlying health conditions, and those with high cumulative socioeconomic and environmental health burdens, are likely to experience more adverse health outcomes when exposed to pollution.[99] Moreover, certain populations, such as populations of color, are disproportionately exposed to air pollution in the United States.[100] Ensuring that procedural equity is prioritized can help all affected communities to have a voice in deciding how to value the tradeoffs among emission types and locations.

---

[96] Katzenstein, W., and J. Apt. 2009. "Air Emissions Due to Wind and Solar Power." *Environmental Science & Technology* 43(2):253–258.

[97] Precise impacts for particular systems require analysis of unit commitment and economic dispatch functions. Mills, A., R. Wiser, M. Milligan, and M. O'Malley. 2009. "Comment on 'Air Emissions Due to Wind and Solar Power.'" *Environmental Science & Technology* 43(15):6106–6107.

[98] Accurate estimates of the distributional emissions impacts of DG adoption—and the value of DG—require grid modeling to capture the operation of peaker power plants, energy storage, and other grid resources. Dispersion modeling would be needed to assess exposure to air pollution for populations. It should be noted that while the health impacts of air pollutant emissions tend to be highest (per capita) near the emission source the cumulative health impacts of stack emissions may extend over broad regions and across state boundaries. NRC (National Research Council). 2010. *Global Sources of Local Pollution: An Assessment of Long-Range Transport of Key Air Pollutants to and from the United States*. Washington, DC: The National Academies Press. https://doi.org/10.17226/12743.

[99] Casey, J.A., J.G. Su, L.R. Henneman, C. Zigler, A.M. Neophytou, R. Catalano, R. Gondalia, Y-T Chen, L. Kaye, S.S. Moyer, V. Combs, G. Simrall, T. Smith, J. Sublett, and M.A. Barrett. 2020. "Coal-Fired Power Plant Closures and Retrofits Reduce Asthma Morbidity in the Local Population." *Nature Energy* 5(5):365–366.

[100] Tessum, C.W., D.A. Paolella, S.E. Chambliss, J.S. Apt, J.D. Hill, and J.D. Marshall. 2021. "PM2.5 Polluters Disproportionately and Systemically Affect People of Color in the United States." *Science Advances* 7(18):eabf4491.

The assignment of value to GHG reductions has equity dimensions as well. The climate impacts of any change in GHGs do not depend on where these changes occur, but their value does depend on how future generations are weighted, relating to intergenerational equity. As described in Chapter 4, the Social Cost of Carbon is used in many cases to calculate the value of emission reductions, but these future impacts are typically discounted; the higher the discount rate, the less the impact on future generations is valued. Furthermore, the assignment of a monetary value to future climate impacts, including damage to the environment and public health, necessarily requires some measure of agreement on the monetary value of non-monetized items such as human life, raising ethical and procedural justice questions regarding the values used and who gets to determine these values.

Household-Level Health Benefits from Clean DG Adoption

The adoption of BTM solar, storage, and other clean DG, supported by net metering, rarely reduces on-site air pollutant emissions, except in the case of solar + storage systems being used to displace a diesel generator used for off-grid electricity or backup power.[101] However, BTM solar may provide indirect human health benefits and/or facilitate cost-effective home electrification, although these may be hard to quantify. Depending on the net metering policy in place, the installation of BTM solar can provide lower and more consistent energy bills for adopters and limit the risk of shutoffs.[102] In addition, the adoption of solar + storage resources can provide resilience in the case of emergencies, further ensuring access to power during critical periods and for elderly, disabled, medical device-dependent, and other vulnerable customers. See Box 5-1 for further information on net metering and resilience. Because these populations frequently face barriers to adoption of clean energy (e.g., access to capital), net metering may require modification or need to be coupled with additional policies to ensure that these additional benefits accrue to those who may most benefit.

[101] Katalenich S.M., and M.Z. Jacobson. 2021. "Renewable Energy and Energy Storage to Offset Diesel Generators at Expeditionary Contingency Bases." *The Journal of Defense Modeling and Simulation* online, https://doi.org/10.1177/15485129211051377; Shakya, S.R., I. Bajracharya, R.A. Vaidya, P. Bhave, A. Sharma, M. Rupakheti, and T.R. Bajracharya. 2022. "Estimation of Air Pollutant Emissions from Captive Diesel Generators and Its Mitigation Potential Through Microgrid and Solar Energy." *Energy Reports* 8:3251–3262.

[102] Hernández, D. 2016. "Understanding 'Energy Insecurity' and Why It Matters to Health." *Social Science & Medicine* 167:1–10; Zanocco, C., J. Flora, R. Rajagopal, and H. Boudet. 2021. "When the Lights Go Out: Californians' Experience with Wildfire-Related Public Safety Power Shutoffs Increases Intention to Adopt Solar and Storage." *Energy Research & Social Science* 79:102183.

BOX 5-1
EQUITY DIMENSIONS OF NET METERING RELATED TO RESILIENCE

In recent years, extreme weather events and growing climate stressors such as hurricanes and wildfires have led to electricity grid outages across the United States lasting for days and even weeks. Moreover, some of the strategies used to mitigate wildfires—such as the Public Safety Power Shutoffs (PSPS) employed in California to de-energize lines and reduce the risk of sparking wildfires—have led to extended grid outages themselves. The impacts of these outages are often inequitable because they pose a particular risk to vulnerable populations, such as those dependent on electricity to power medical devices or refrigerate medicines, elderly populations at risk of heat stroke without sufficient access to air conditioning, or those with underlying medical conditions.

To mitigate both the health risks and everyday inconveniences posed by extended grid outages, increasing numbers of households have begun to adopt combined solar and energy storage systems with the capability to island from the grid and provide backup power.[a] Much like the trends seen for rooftop solar adoption,[b] however, energy storage adoption is more widespread in wealthier, whiter communities in California,[c] and similar trends are likely elsewhere. Disabled, elderly, low-income customers, and others facing high socioeconomic and public health burdens may be the ones most in need of resilient power systems, but they are least likely to have access.

Community-level resilience solutions, such as virtual net metered solar and storage-powered community resilience hubs,[d] can address some of these equity issues by providing places for vulnerable populations to access cooling, breathe filtered air, refrigerate medicines, charge cell phones, and obtain other critical services.[e] The economic feasibility of these systems, however, is determined in part by the interplay of available support and incentives, including net metering policy, for both solar *and* energy storage.

In some cases, the lowest-cost resilient design entails oversizing solar systems relative to typical demand to ensure that these systems can provide sufficient power to recharge batteries through multi-day outages, rather than installing a battery system that can last for multiple days on its own. Current net metering policies, however, may compensate customers at a much lower rate for any generation in excess of net annual demand, meaning oversized solar systems for resilience benefits will receive minimal compensation for their generation. It also may be beneficial to link solar installations on multiple buildings—on multiple meters—to operate as part of a microgrid in the case of an outage, but this is not allowed under net metering rules in most jurisdictions. California is currently exploring regulatory mechanisms to enable neighboring buildings on separate meters to island together as a small microgrid.[f]

Solar + storage systems can also provide resilience benefits to individual households and may be particularly beneficial in historically underserved and disadvantaged communities. However, net metering is not always designed to achieve resilience goals; traditional net metering and its variants may need to be modified and/or combined with complementary incentives achieve these goals.

The impact of net metering rates on the deployment of resilient solar + storage, particularly for vulnerable communities, will depend on:

- The level of compensation for any excess solar generation when coupled with a resilient storage system
- The availability of targeted incentives or net metering designed to increase access for vulnerable populations
- Regulatory design affecting the ability of multiple net-metered systems to island together in the case of utility grid disturbances
- The availability of complementary incentives to support resilience goals

---

[a] A moderate amount of resilience can also be provided by rooftop solar systems even without storage, as long as those systems are designed to island from the grid with the inclusion of a separate disconnect in the case of emergencies. In this situation, a separate outlet is often connected directly to the inverter, providing power while the sun is shining but not at night. During grid outages, most behind-the-meter solar photovoltaic owners do not have any access to power from their panels unless they have storage installed.

[b] Lukanov, B.R., and E.M. Krieger. 2019. "Distributed Solar and Environmental Justice: Exploring the Demographic and Socio-Economic Trends of Residential PV Adoption in California." *Energy Policy* 134:110935; Reames, T.G. 2020. "Distributional Disparities in Residential Rooftop Solar Potential and Penetration in Four Cities in the United States." *Energy Research & Social Science* 69:101612; Sunter, D.A., S. Castellanos, and D.M. Kammen. 2019. "Disparities in Rooftop Photovoltaics Deployment in the United States by Race and Ethnicity." *Nature Sustainability* 2(1):71–76.

[c] Brown, D.P. 2022. "Socioeconomic and Demographic Disparities in Residential Battery Storage Adoption: Evidence from California." *Energy Policy* 164:112877.

[d] Incentives such as California's Self Generation Incentive Program (SGIP) can support the adoption of energy storage for resilience, and SGIP includes carve-outs for disadvantaged communities, low-income households, households in wildfire zones, those who have faced multiple public safety power shutoffs in recent years, households with electric water pumps, and customers needing power for medical devices CPUC. 2021d. "Self-Generation Incentive Program (SGIP)." https://www.cpuc.ca.gov/industries-and-topics/electrical-energy/demand-side-management/self-generation-incentive-program.

[e] Baja, K. 2018. "Resilience Hubs: Shifting Power to Communities and Increasing Community Capacity." Urban Sustainability Directors Network (USDN). https://www.usdn.org/uploads/cms/documents/usdn_resiliencehubs_2018.pdf.

[f] CPUC. 2022e. "Resiliency and Microgrids." https://www.cpuc.ca.gov/resiliencyandmicrogrids.

CURRENT STATE OF EQUITY REQUIREMENTS AND RANGE OF EQUITY SOLUTIONS

The following conclusions from preceding sections set the stage for assessing opportunities for net metering to enhance distributional, procedural, intergenerational, and structural equity with respect to BTM investments:

- Customers who currently adopt BTM technologies and participate in net metering tend to be higher-educated, wealthy, white, and homeowners, and they are clustered in particular states and regions of the country.[103,104]
- Barriers exist for various constituencies to access BTM technologies and therefore associated net metering support, such as limited disposable income, savings, access to financing for low-income households, landlord-tenant problems for renters, and lack of access to smart meters and broadband internet.
- More costs are shifted to non-participants from net metering participants when there are high levels of BTM DG. When societal impacts such as resilience, health, and various environmental benefits of BTM DG are not considered, the benefits may outweigh the added costs.
- When societal impacts are considered (such as resilience, reliability, clean air, carbon mitigation, and health), there may be no net cost shift between net metering participants and non-participants, although there may still be rate impacts on non-participants.

Thus, there are many significant equity implications for the design of net metering, the role of net metering variants (which relate to rates), net metering alternatives (which are external to rates), and other public policies (including taxpayer-funded programs) to address access and investment in BTM technologies and net metering.

Opportunities to Enhance Distributional Equity

No single approach provides a perfect solution to enhancing the distributional equity of net metering, and each approach poses new and unique challenges. Improving distributional equity requires strategies to ensure previously underserved and structurally disadvantaged populations receive benefits from net metering while simultaneously mitigating any net metering impacts, such as possible cost shifts, to non-participants.

[103] Forrester, S., G. Barbose, E. O'Shaughnessy, N. Darghouth, and C.C. Montañés. 2022. *Residential Solar-Adopter Income and Demographic Trends: November 2022 Update*. Berkeley, CA: Lawrence Berkeley National Laboratory.

[104] O'Shaughnessy, E., G. Barbose, R. Wiser, S. Forrester, and N. Darghouth. 2020. "The Impact of Policies and Business Models on Income Equity in Rooftop Solar Adoption." *Nature Energy* 6(1):84–91.

An optimal approach may include several elements. For instance, net metering could be structured as buy-all and sell-all agreements, where clear communication of the utility's commitment to more efficient rate design could encourage complementary investments that create new sources of value and customer savings.[105]

Further research and pilot rate implementation are required to understand the tradeoffs between different net metering and rate design approaches in terms of equity and efficiency. For example,

- LMI customers and households in historically underserved or vulnerable communities could remain eligible for traditional net metering while all other households are shifted onto net billing or other alternative rates. Such an approach would have the added benefit of ensuring that LMI households receive the same financial benefits from PV adoption as early high-income adopters received.
- Buy-all and sell-all agreements could be structured with higher sell-all rates for LMI customers, or such that customers receiving bill assistance are compensated for solar generation at retail rates rather than discounted bill assistance rates. For example, in Mississippi's version of net billing, LMI customers receive a 2.5 cent/kWh adder over avoided cost for excess generation compensation, plus an additional 2.0 cents/kWh for households with annual incomes up to 225 percent of the federal poverty level.[106]
- LMI net metering customers could be exempt from fixed charges, minimum bills, or demand charges. For example, Virginia has a minimum bill for community solar participants, but low-income customers are exempt from it.[107]

Programs and incentives could also be offered outside of net metering to support rooftop solar adoption by LMI customers.[108]

- California's Single Family Affordable Solar Homes program bought down the upfront costs of solar for income-qualified customers,[109] and its multifamily counterpart the Solar on Multifamily Affordable Housing (SOMAH) program

---

[105] Burger, S., I. Schneider, A. Botterud, and I. Pérez-Arriaga. 2018. *Fair, Equitable, and Efficient Tariffs in the Presence of Distributed Energy Resources.* Cambridge, MA: MIT Center for Energy and Environmental Policy Research.

[106] For further information on Mississippi's net metering program, see https://programs.dsireusa.org/system/program/detail/5841.

[107] Virginia Electric Utility Regulation Act, Title 5, § 56-594.3. Shared Solar Programs.

[108] DOE. 2021. "Solar Futures Study." https://www.energy.gov/eere/solar/solar-futures-study.

[109] CPUC. 2022b. "CSI Single-Family Affordable Solar Homes (SASH) Program." https://www.cpuc.ca.gov/industries-and-topics/electrical-energy/demand-side-management/california-solar-initiative/csi-single-family-affordable-solar-homes-program.

offers bill credits to affordable housing tenants who have solar installed on their building.[110]

- California's Self Generation Incentive Program (SGIP) offers incentives to LMI and medical baseline customers (those who rely on electricity to power medical equipment) who install energy storage in their homes. Higher incentives are given to vulnerable customers who are in regions with high risk of wildfires and power shut-offs to help ensure these households and communities can remain resilient during outages.[111]
- Low-income households could be exempted from paying interconnection or application fees for distributed solar.[112]
- Virtual net metering and community solar projects can provide carve-outs for LMI customers.

Net metering alternatives that rely on non-ratepayer funds (e.g., tax revenues) could be used to support solar adoption for target populations. As noted by Mathew Freedman in a presentation to the committee, tax revenues are more progressive in their collection and can be used to offset the need for subsidy sources for retail rates.[113] For example, revenues could provide low- or no-interest financing or upfront incentives for customers in LMI and disadvantaged communities. Incentives provided from sources other than utility costs, such as tax revenues, have a less regressive impact and should be considered to address equity.[114] Efforts to identify revenues that are progressive in their collection should be used whenever possible to avoid disproportionate rate increases for low-income non-participants.

Addressing Equity in Regulatory Frameworks

As of the time of writing, 36 jurisdictions have recently been charged with the responsibility to consider several new transformational values that are relevant to the design of

[110] CPUC. 2021e. "The Solar on Multifamily Affordable Housing (SOMAH) Program." https://www.cpuc.ca.gov/industries-and-topics/electrical-energy/demand-side-management/somah.

[111] CPUC. 2021d. "Self-Generation Incentive Program (SGIP)." https://www.cpuc.ca.gov/industries-and-topics/electrical-energy/demand-side-management/self-generation-incentive-program.

[112] Brown, M.A. 2022. "Testimony Before the Georgia Public Service Commission on Georgia Power's 2022 Rate Case." Document Filing 192146. Georgia Public Service Commission. https://psc.ga.gov/search/facts-document/?documentId=192146.

[113] Presentation to the committee on "The Role of Net Metering in the Evolving Electricity Systems." National Academies of Sciences, Engineering, and Medicine on August 9, 2022.

[114] Ibid.

equitable net metering tariffs and other energy regulations, policies, and programs.[115] Many of these new authorities pertain to environmental benefits and costs and equity goals that can now be considered when designing and adjudicating net metering (and other policies). For example:

- Fifteen states have begun mapping locations of particular focus for equity and environmental justice.
- Many states will be addressing conditions in "disadvantaged," "disproportionately impacted," "environmental justice," "environmental justice populations," "native tribes," or "frontline" communities (including California, Colorado, Illinois, Maine, Maryland, New Jersey, New Mexico, New York, and Washington).
- Seventeen states have specified addressing equity values using terms such as diversity, equity, inclusion, environmental equity, and environmental justice (Alabama, California, Connecticut, Delaware, Illinois, Maine, Maryland, Massachusetts, Michigan, Minnesota, New Jersey, New York, North Carolina, Oregon, Rhode Island, Washington, and Wisconsin).
- Several states have new authorities to specify distributional goals, that are generally consistent with Justice40 goals including a minimum of 35 percent of benefits from clean energy and energy efficiency programs accruing to

[115] Sources*:* All web pages retrieved by Tom Stanton in May 2022. See CESA. n.d. "Table of 100% Clean Energy States." https://www.cesa.org/projects/100-clean-energy-collaborative/guide/table-of-100-clean-energy-states; Climate Xchange. 2022. "State Climate Policy Tracker." https://climate-xchange.org/network/map; Climate Xchange. 2022. "State Climate Policy Tracker (Spreadsheet), Version 1.3." https://docs.google.com/spreadsheets/d/1t912_uR-x8DeMTDNCm6y9nb6QLiiLDAdWNciteu1rOU/edit#gid=951166054; Colorado State University, Center for the New Energy Economy, and The Nature Conservancy. "Methodology Behind the State Policy Opportunity Tracker (SPOT) for Clean Energy." https://spotforcleanenergy.org/about; E9 Insight. 2021. "Pathways to Changing the PUC Mandate: A Regulatory Review." Prepared on behalf of the Institute for Market Transformation. https://drive.google.com/file/d/1c244_ryG4C15vnG1tuGu_1m8wgn9DwnK/view; E9 Insight. 2021. "PUC Mandate Database—Final." Prepared on behalf of the Institute for Market Transformation. https://docs.google.com/spreadsheets/d/1acwblw_Q67d_bpdKxFQzJbsvP19WGj59/edit#gid=286574870; Klee, R.J., and S. Baldinger. 2021. "Review of State Public Utility Commission Statutory Mandates." Report for the Institute for Market Transformation PUC Mandate Project. New Haven, CT: Center for Business and the Environment at Yale. https://drive.google.com/file/d/1bvQTifqgxGS5i7_CU1uj0HLPBQW5kjYV/view; NRRI. 2022b. "State Clean Energy Policy Tracker." https://www.naruc.org/nrri/nrri-activities/clean-energy-tracker; Zitelman, K., and J. McAdams. 2021. *The Role of State Utility Regulators in a Just and Reasonable Energy Transition—Examining Regulatory Approaches to the Economic Impacts of Coal Retirements.* Washington, DC: National Association of Regulatory Utility Commissioners. https://pubs.naruc.org/pub/952CF0F2-1866-DAAC-99FB-0C6352BF7CB0.

disadvantaged communities (New York),[116] and creating a Justice40 Oversight Committee (Delaware).[117,118]

Improved modeling based on new data for equity mapping can make it possible to better evaluate the equity of net metering ratemaking variants and alternative outside rates. Vulnerable communities and individuals are hidden in aggregated metrics; most studies do not have sufficient levels of resolution. As experience with equity mapping grows, alternative net metering design features may result. For instance, it may be possible to design geographically based rates for exported generation that are tagged to the time-dependent capacity available on specific circuits. These efforts will only be feasible, however, if data collection is sufficient and transparently shared with stakeholders.

Opportunities to Enhance Net Metering Procedural Equity

Planning for net metering and other tariff designs needs to engage the views of a broader spectrum of stakeholders just as integrated resource planning and distribution system planning processes have begun to do. Currently, community participation in planning is often limited and biased toward well-resourced groups that can retain legal counsel and have the capacity to work with the complexities of rate-case litigation. While some agencies provide utility-funded intervenor compensation, which can enable under-resourced groups to participate in proceedings, such engagement can still be prohibitively complicated.

One way to address procedural equity is to create opportunities for dialogue with broad representation. Development of alternative net metering design features in

[116] NYSERDA (New York State Energy Research and Development Authority). 2022a. "New York State Releases Draft Disadvantaged Communities Criteria to Advance Climate Justice." March 9, 2022. https://www.nyserda.ny.gov/About/Newsroom/2022-Announcements/2022-03-09-New-York-State-Releases-Draft-Disadvantaged-Communities-Criteria.

[117] Reta, M. 2023. "How States Can Help Implement the Justice40 Initiative." *NRDC Expert Blog*, January 27. https://www.nrdc.org/experts/mikyla-reta/how-states-can-help-implement-justice40-initiative.

[118] The Biden-Harris administration created the Justice40 Initiative to confront and address decades of underinvestment in disadvantaged communities. The initiative will bring resources to communities most impacted by climate change, pollution, and environmental hazards. For more information see https://www.transportation.gov/equity-Justice40.

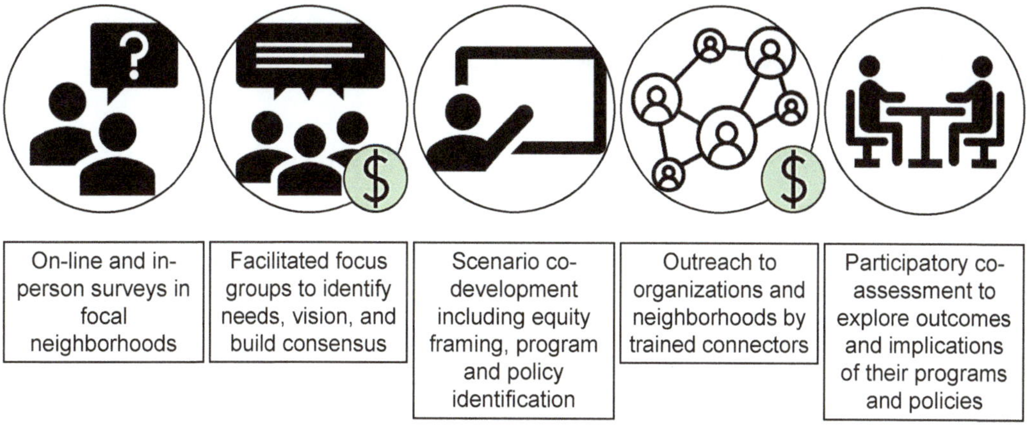

FIGURE 5-3 Typical steps in knowledge co-production to enhance procedural equity.
NOTE: Dollar designs indicate steps where incentives to participants should be considered due to the significant effort which may be needed for success.
SOURCE: Informed by Lisska et al. 2021. "Equitable Transit Oriented Development." Transportation Research Board Committee Meeting. June 8, 2022. https://www.nationalacademies.org/event/06-08-2022/data-metrics-and-analytic-methods-for-assessing-equity-impacts-of-surface-transportation-funding-programs-meeting-6.

collaboration, consistent with current best practices for community engagement,[119] can be used to integrate the diverse perspectives from engineers and scientists, local communities, and business and management practitioners,[120] engaging and empowering individuals and communities who have been marginalized.[121] Such approaches lead to more accurate and useful research results, better usability of data, and increased buy-in

[119] An excellent example of this is the work of the Texas State Department of Transportation's planning of alternative transportation modalities. Lisska et al. 2021. "Equitable Transit Oriented Development." Transportation Research Board Committee Meeting. June 8, 2022. https://www.nationalacademies.org/event/06-08-2022/data-metrics-and-analytic-methods-for-assessing-equity-impacts-of-surface-transportation-funding-programs-meeting-6.

[120] Campbell, H., and D. Vanderhoven. 2016. *Knowledge That Matters: Realising the Potential of Co-Production*. Manchester, UK: N8 Research Partnership; Sanchez Rodriguez, R., D. Ürge-Vorsatz, and A.S. Barau. 2018. "Sustainable Development Goals and Climate Change Adaptation in Cities." *Nature Climate Change* 8:181–183.

[121] Bixler, P., J.A. Belaire, K.M. Faust, M. Scoggins, and A. González. 2022. "Exploring the Connection Between Transdisciplinary Co-Production and Urban Sustainability Solutions: A Case Study at an Urban Stream Management Symposium." *Urban Ecosystems* 25:1207–1216; Cook, E.M., M. Berbés-Blázquez, L.M. Mannetti, N.B. Grimm, D.M. Iwaniec, and T.A. Muñoz-Erickson. 2021. "Setting the Stage for Co-Production." Pp. 99–111 in *Resilient Urban Futures, The Urban Book Series*. Cham: Springer; Norström, A.V., C. Cvitanovic, and M.F. Löf. 2020. "Principles for Knowledge Co-Production in Sustainability Research." *Nature Sustainability* 3:182–190.

to policy solutions that are people-centered and user-led.[122] Some of the typical steps in knowledge co-production processes are shown in Figure 5-3.

Opportunities to Address Intergenerational and Structural Equity

Low-income and other distributional inequities associated with BTM DG and net metering are rooted in historical structural inequities, such as underinvestment in infrastructure, redlined mortgage financing, and low levels of home ownership. Broad initiatives to address these inequities may help enable more equitable adoption of DG and participation in net metering.[123] Programs can provide incentives for LMI households living in single- or multi-family homes to install solar. For example, the Solar Massachusetts Renewable Target (SMART) Program gives special preferences to low-income customers.[124] Certain non-ratepayer incentives for solar adoption are not always supportive of LMI participation. Incentives such as point-of-sale rebates, or financing for those with low or no credit, can better enable additional LMI customers to adopt BTM DG and participate in net metering than tax rebates that tend to benefit wealthier customers who have the tax liability to use them. The new Inflation Reduction Act (IRA) has included $27 billion in funding to inject into state, local, tribal, and non-profit entities to provide financing assistance for a variety of GHG reduction projects, with a significant portion of that funding required to benefit low-income and disadvantaged communities.[125] This federal funding will no doubt include projects aimed at helping LMI households install or otherwise have access to solar system resources.

Historic disinvestment in infrastructure needed to support DG adoption in many LMI communities and communities of color can lead to lower grid hosting capacity. Customers may also need to upgrade roofs and other infrastructure before installing solar. Even lack of access to energy efficiency measures may increase energy bills for these

[122] Audia, C., F. Berkhout, G. Owusu, Z. Quayyum, and S. Agyei-Mensah. 2021. "Loops and Building Blocks: A Knowledge Co-Production Framework for Equitable Urban Health." *Journal of Urban Health* 98:394–403.

[123] EPA Administrator Michael S. Regan, has committed the U.S. Environmental Protection Agency to create a national Office of Environmental Justice and External Civil Rights to consider how communities have been impacted over time, considering how counties that are poor in income, health, and political power, and are mostly minorities, have been targeted for environmentally toxic storage tanks, coal plants and their combustion residues, and other environmental injustices. The Biden administration has called on all federal agencies to ensure disadvantaged communities receive 40 percent of the benefits from federal investment in a range of areas including renewable energy.

[124] Lane, C. 2023. "Massachusetts SMART Program Explained." *Solar Reviews*, January 11. https://www.solarreviews.com/blog/massachusetts-smart-program-replaces-srecs.

[125] For additional information see EPA. 2023. "Greenhouse Gas Reduction Fund," https://www.epa.gov/inflation-reduction-act/greenhouse-gas-reduction-fund.

households, which will then require larger solar systems and more capital. Investments, such as those enabled through the IRA, in expanding energy efficiency, fixing rooftops, and upgrading electric grids, could help reduce the barriers for households in historically underinvested communities to adopt DG and participate in net metering.[126]

FINDINGS

All electricity rates, including net metering tariffs, can raise significant equity issues. These equity concerns exist across a variety of dimensions: *distributional equity*, including who receives net metering benefits and what impacts net metering has on energy burdens, energy insecurity, and energy poverty considerations; *procedural equity*, related to whether diverse voices are included in the process of setting rates; *structural equity*, reflecting issues such as barriers to net metering introduced by historic disinvestment in infrastructure; and *intergenerational equity*, involving considerations that arise when considering long-term impact of decisions, such as those regarding climate change.

To produce more equitable outcomes, policymakers need to build equity considerations into the design of net metering, its variants, and alternatives, inclusive of companion policies such as legislative and taxpayer-funded programs. The committee suggests that all four dimensions of equity (distributional, procedural, structural, and intergenerational) serve as guideposts for all utility rate and program designs, including net metering. The committee's specific findings and recommendations address the distribution of net metering benefits among customers; potential adverse impacts of net metering, such as cost shifts; and additional direct and indirect societal considerations associated with net metering.

Finding 5-1: Any rate and bill impacts resulting from net-metered BTM DG may accrue disproportionately to low-income and other disadvantaged households that have faced barriers to solar adoption. Low-income and energy-burdened households may be less able than other customers to bear any bill increases if there are cost shifts.

Finding 5-2: BTM DG, which can be enabled by net metering, holds the potential to both reduce costs and provide bill stability, which has particular value for populations

[126] Advocacy groups have also launched community-based solar programs to enable low-income households to enjoy the benefits of distributed rooftop generation, such as buying shares of community solar systems. Brown, M.A., A. Soni, M.V. Lapsa, K.A. Southworth, and M. Cox. 2020b. "High Energy Burden and Low-Income Energy Affordability: Conclusions from a Literature Review." *Progress in Energy* 2(4):1–35; Weissman, G. 2017. "Rental Rooftops: A Missed Opportunity for Solar Energy." https://frontiergroup.org/blogs/blog/fg/rental-rooftops-missed-opportunity-solar-energy.

facing high energy cost burdens and energy insecurity, such as low-income households, renters, populations of color, the elderly, and those dependent on electric power for operating home medical care devices.

Finding 5-3: Barriers, such as lack of access to financing, hinder BTM DG adoption and net metering participation for certain populations (low-income households, renters, and communities of color). As a result, they have not benefited from net metering or from other BTM policies to the same degree as their wealthier, White, and home-owning counterparts.

Finding 5-4: The trends show that both inclusive financing techniques and community solar are helping to enable lower-income households to access the benefits of renewable energy and electricity bill savings in more meaningful numbers.

Finding 5-5: Community solar programs provide an opportunity for renters, homeowners with a shaded or poorly maintained roof, and households without access to capital to receive some of the benefits of distributed solar.

Finding 5-6: Net metering tariffs or complementary incentives can be designed to reflect locationally specific benefits of DG, such as grid resilience in vulnerable communities, the deferral of grid investments, and to support pollution reduction goals in environmentally overburdened areas.

Finding 5-7: Net metering can be used to support the installation of solar + storage to achieve resilience goals, including household resilience, community resilience, and grid resilience. All of these may be of particular benefit for certain vulnerable populations such as the elderly and medical baseline customers. However, net metering tariffs often limit the ability to adopt potentially cost-effective resilient BTM project designs, such as over-sizing solar systems or linking multiple rooftops. Recognizing that oversizing or linking of systems could introduce more direct federal involvement over the pricing of exports from such BTM systems, as discussed in Chapter 7, net metering tariffs might be designed to enable such outcomes or be developed in parallel with companion policies, particularly for disadvantaged customers.

Finding 5-8: Customers seeking to determine whether to adopt BTM DG and participate in net metering need access to basic information about system costs and utility bill impacts in a form that they can understand and use. Such information will help to equip customers with information to make decisions, thus enhancing procedural equity.

Finding 5-9: Experienced regulatory practitioners and intervenors in rate cases need to have access to much more information, to determine whether rates under net metering policy reflect cost-of-service for different customer classes, the cost of displaced electricity, the value of deferring the construction of new distribution, transmission, and generation (particularly fossil fuel plants).

RECOMMENDATIONS

Recommendation 5-1: Rates should be designed consistent with updated ratemaking principles, with particular attention to the equity impacts for customers least able to afford them. To help accomplish this, utilities and policymakers should ensure that information about utility rates is easily available to all customers (e.g., not just through postings on websites), and that all customers have a voice and can participate in the design of rates.

Recommendation 5-2: Decision makers on electricity rates should consider both the impacts of the distribution of benefits and costs, as well as total benefits and costs when designing net metering policies, and ensure that adequate data are collected and made publicly available to do so. These benefits and costs should include and balance, among other things: public health impacts, job impacts, land use impacts, and the future options that will be enabled or precluded.

Recommendation 5-3: Fair ratemaking should allow for affected consumers to have a meaningful voice in the process (so as to ensure procedural equity) and take into consideration (among other things) the potential impacts of rates on various customer segments (from an economic efficiency and equity point of view), and on customers' decisions about adopting BTM technologies or load or grid defection in the context of technology trends. Regulators and decision makers of publicly owned utilities should ensure that such information is provided to the public and is easily understandable and accessible.

Recommendation 5-4: Investments to reverse historic structural inequities, such as grid upgrades in historically disinvested communities to increase hosting capacity,

should be considered to enable broader adoption of BTM DG and participation in net metering.

Recommendation 5-5: Policymakers should consider expanding community and government-funded programs, such as low- or no-interest financing, to expand access to BTM DG by low- and medium-income customers and their participation in net metering. In addition, incentives that focus on income-qualified customers and renters can help reduce inequities due to lack of access.

Net Metering and Distributed Energy Technologies

Prior chapters of this report examined net metering and distributed generation from various angles. This chapter lays out the technology implications of increasing amounts of behind-the-meter (BTM) distributed generation (DG) for the larger electricity system and describes the ongoing technological advances that enable the integration of BTM DG/distributed energy resources (DER) into the electricity grid, followed by a discussion of new technologies that are needed to ensure that the integration of BTM DG/DER leads to a reliable and resilient grid. The chapter underscores the technical ability of BTM DG supported by net metering to provide tools not only to benefit the DG customer but also to benefit the grid at large, in terms of both reliability and resilience.

Briefly, this chapter provides an overview of the effect on the grid of BTM DG and related technologies influenced by net metering and supported by grid technologies to integrate, manage, or control them. It presents a discussion of:

- Grid modernization and the integration of renewables
- Power-physics-based constraints emerging from integrating BTM DG onto the grid
- Emerging advances in power electronics, storage, communications, and control technologies
- Cybersecurity and resilience considerations emerging from BTM DG and net metering
- Chapter summary and findings

GRID MODERNIZATION AND THE INTEGRATION OF RENEWABLES

Grid modernization[1] is introducing innovative technologies and services to the electricity grid at an unprecedented rate. The focus of this chapter is distributed renewable

[1] Corresponding with the Department of Energy's Grid Modernization Initiative (GMI), grid modernization can be defined as the development of the "concepts, tools, and technologies needed to measure, analyze, predict, protect, and control the grid of the future," including those which "help integrate all sources of electricity better, improve the security of our nation's grid, solve challenges of energy storage and distributed generation, and provide a critical platform for U.S. competitiveness and innovation in a global energy economy." See DOE. 2022. "Grid Modernization Initiative." https://www.energy.gov/gmi/grid-modernization-initiative.

sources of electricity such as solar, storage such as batteries, and demand response such as management of variable load. Collectively these technologies are denoted as DER; distributed generation is denoted as DG. A good portion of DG is BTM and is made up of rooftop or on-site solar. BTM DG poses a challenge to traditional top-down utility planning and grid operation, as they inject uncertainty into the electricity grid operations due to their variability, uncertainty, and location of adoption. They also pose challenges for traditional regulatory structures and utility business models.[2]

A shift in generation toward variable renewable energy, such as wind and solar, is a shift away from the traditional fossil-based generation that is dispatchable or readily available. The shift to cleaner and more distributed resources may reduce overall societal and system costs, taking into account the full costs of externalities, including greenhouse gas (GHG) emissions, pollution, and health impacts. However, addressing the challenge of decentralization and variability increases the cost of adding renewable electricity to the power grid due to investments in advanced communication, control, protection, and resilience needed to ensure an uninterrupted supply of generation to meet customer demand and for the grid to function reliably. As the share of renewables grows, which will be essential for deep decarbonization,[3] accounting for and allowing recovery of these system-wide costs of integrating variable renewable resources will require regulatory attention (see Chapter 7) and possibly regulatory and utility business model innovation.

The following sections expand on the technology implications of increasing amounts of BTM DG for the larger electricity system, the ongoing technological advances that enable the integration of BTM DG/DER into the electricity grid, and new technologies that are needed to ensure that the integration of BTM DG and DER leads to a reliable and resilient grid.

TECHNOLOGY INTERSECTION 1: POWER PHYSICS-BASED CONSTRAINTS

BTM DG capacity is expected to increase significantly at the grid edge[4] in the coming years, influenced by net metering compensation and other policies. Such DG and other DER may include increasing installation of solar and wind resources, stationary storage, mobile resources such as electric vehicles, and flexible consumption combined

[2] NARUC. 2022. "Task Force Cohort Roadmaps." https://www.naruc.org/taskforce/resources-for-action/roadmaps.

[3] NASEM (National Academies of Sciences, Engineering, and Medicine). 2021. *Accelerating Decarbonization of the U.S. Energy System*. Washington, DC: The National Academies Press. https://doi.org/10.17226/25932.

[4] Mai, T.T., P.H. Nguyen, Q-T Tran, A. Cagnano, G. De Carne, Y. Amirat, A-T. Le, and E. De Tuglie. 2021. "An Overview of Grid-Edge Control with the Digital Transformation." *Electrical Engineering* 103:1989–2007.

with storage, enabled by advanced demand response or demand management tools. To the extent that variable DG capacity, such as solar and wind, grows, the challenges of maintaining the real-time balance of supply and demand on the system and the stable operation of power systems could increase. New economic models and distribution regulation technologies and tools will likely be required to address these challenges.

BTM DG Integration

The increase of DG is also significantly affecting the flow of electricity on both sides of the meter.[5] The existing framework of power delivery and associated grid services, such as voltage control, frequency control, power quality, and outage management, are all affected by the changing technical conditions and requirements, which affect the underlying costs to deliver electricity in a reliable manner during normal operation and in a resilient manner following outages and disruptions.

While customers with BTM DG have the technological capability to provide valuable energy and flexibility services to the distribution grid, this DG needs to be integrated into the grid, not just interconnected to it.[6] Successful BTM DG integration at high penetration rates and in ways that enable these resources to provide grid services will depend on addressing the operations challenges of the local distribution grid. In the absence of appropriate communications, control and other technologies, and a regulatory framework that supports investment in them, DG has the potential to disrupt distribution grid operation through reverse power flows, voltage violations, harmonics, phase unbalance power quality concerns, the mal/mis-operation of protection devices, as well as presenting grid bottlenecks and congestion.[7] Avoiding these problematic outcomes is essential (see Box 6-1 for one example of how DG systems may support system security).

The following itemizes some of the main technological challenges in BTM DG integration. The first challenge due to increased BTM DG is *ensuring the real-time balance of supply and demand.* With increased rooftop solar, the net injection from and into the

---

[5] Erdener, B.C., C. Feng, K. Doubleday, A. Florita, and B.M. Hodge. 2022. "A Review of Behind-the-Meter Solar Forecasting." *Renewable and Sustainable Energy Reviews* 160:112224; Revuelta, P.S., P.L. Salvador, and J.P. Thomas. 2016. "8 - Distributed Generation," pp. 285–322 in *Active Power Line Conditioners: Design, Stimulation and Implementation for Improving Power Quality*. Amsterdam: Elsevier Inc.

[6] In theory, a cluster of microgrids and universal (or near universal) BTM DG, working together with storage, have the potential to provide a far more technically resilient grid than the current mostly centralized generation model used today. However, unlocking the value of that theoretical possibility depends on a number of enabling conditions, some of which are discussed in this chapter.

[7] Yazdani, H., M. Doostizadeh, and F. Aminifar. 2022. "Unlocking the Value of Flexibility of Behind-the-Meter Prosumers: An Overview of Mechanisms to Esteemed Trends." *The Electricity Journal* 35(5):107126.

BOX 6-1
INTEGRATING HYBRID DG+BATTERY+GENERATOR FOR SYSTEM SECURITY

To provide resilience with large amounts of BTM DG and battery storage in more challenging environments a more complex system may be needed, made up of photovoltaics (PV)+ battery+cogeneration (or a combined heat and power [CHP] unit) with integrating technologies as shown in Figure 6-1 below.

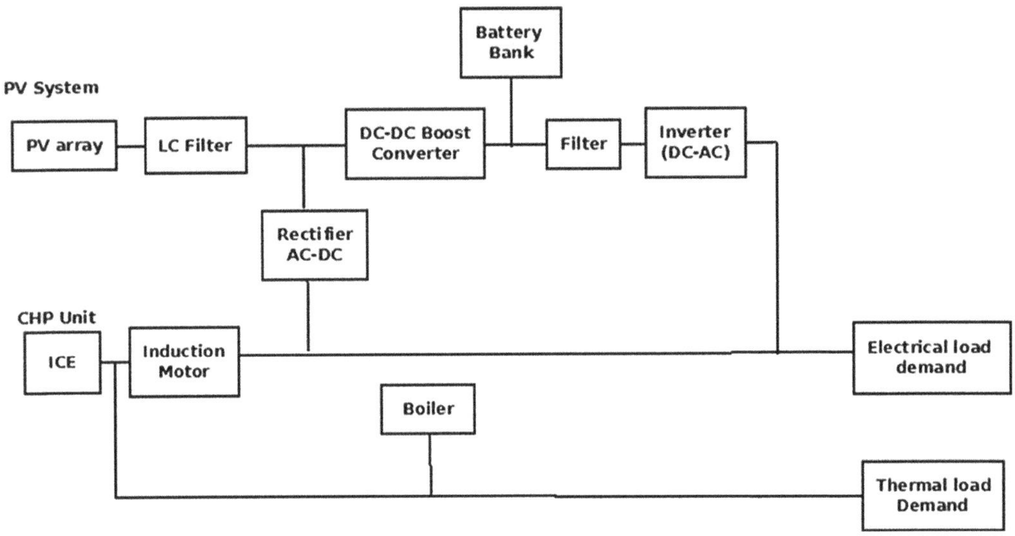

FIGURE 6-1 Block diagram of PV + CHP + battery system.
SOURCE: Reprinted from K.K. Shah, A.S. Mundada, and J.M. Pearce, 2015, "Performance of US Hybrid Distributed Energy Systems: Solar Photovoltaic, Battery and Combined Heat and Power," *Energy Conversion and Management* 105:71–80, https://doi.org/10.1016/j.enconman.2015.07.048. Copyright 2015, with permission from Elsevier.

grid varies significantly with location and time (e.g., time of day, season of the year, and variability in cloud cover). As a result, ensuring that power is balanced, that is, where power generated equals power consumed, becomes technically complex, especially since this balance has to be ensured at all times and at all locations. If sufficient storage capacity is not available to address any imbalance, the challenge will be to develop and adopt alternative technologies that address these imbalances and/or pursue program and rate designs that can do so. For example, Hawaiian Electric's Smart Export Program requires distributed solar systems to be paired with battery storage and only provides compensation for electricity exported during the hours of 4:00 pm to 9:00 am.[8]

[8] For more information see https://www.hawaiianelectric.com/products-and-services/customer-renewable-programs/rooftop-solar/smart-export.

In addition to addressing imbalance issues and reducing reliance on conventional backup generation from the grid as the integration of distributed (and grid-scale) wind and solar increases, storage can also provide benefits to conventional electric power generation by reducing the need for fossil-fuel-based power plants that only run at times of peak load. The traditional storage technology that has been used to address the challenge of variability is pumped storage hydropower (pumped hydro).[9] Over the past few years, the technological landscape of storage has been changing rapidly, with battery storage as one of the DER options increasingly available to manage the variability of renewable energy.[10] Storage is now economical at a level where it is being deployed in nearly every utility service territory. In addition, advances are being reported in multiple kinds of storage capability, including thermal, chemical, and mechanical energy.[11] Encouraged by these technological advances, the improving economics of storage, and flexible loads that can provide grid support at an equal or lower cost than peaking power plants, utilities in many states are already starting to procure significant amounts of storage to address variability and balance real-time supply and demand.[12]

The second technological challenge is *managing imbalances across the three phases (or wires)*[13] *of alternating current* used to deliver power to customer locations. Imbalance occurs whenever loads (e.g., new houses) are added, which are often connected to the grid through a single phase (see Box 6-1). Imbalance can also occur with variable generation, such as BTM solar. The imbalance can grow with increased BTM DG, because not only does it cause two-way flows but two-way flows that keep changing. In other words, one customer could be buying from the grid while the other is selling, which can be difficult

[9] Pumped-hydro storage units are designed to pump water into a reservoir during times of low electricity demand and prices, and the stored water to be released for power generation at times of high demand and prices. However, these systems are very site specific and can be expensive. See EERE. 2022c. "What Is Pumped Storage Hydropower?" https://www.energy.gov/eere/water/pumped-storage-hydropower.

[10] EIA. 2021a. *Battery Storage in the United States: An Update on Market Trends*, pp. 5–12. Washington, DC: Department of Energy.

[11] Massachusetts Institute of Technology Energy Initiative. 2022. *The Future of Energy Storage*. Cambridge, MA: Massachusetts Institute of Technology. https://energy.mit.edu/publication/the-future-of-energy-storage.

[12] Center for Sustainable Systems. 2021. *U.S. Grid Energy Storage Factsheet*. No. CSS15-17. Ann Arbor, MI: University of Michigan. https://css.umich.edu/publications/factsheets/energy/us-grid-energy-storage-factsheet; DOE Global Energy Storage Database. 2021. "Search Projects." https://sandia.gov/ess-ssl/gesdb/public/projects.html; Edison Electric Institute. 2021. "Leading the Way: U.S. Electric Company Investment and Innovation in Energy Storage." https://www.eei.org/-/media/Project/EEI/Documents/Issues-and-Policy/Energy-Storage/Energy_Storage_Case_Studies-062021.pdf; SEPA (Smart Electric Power Alliance). 2019. "2019 Utility Energy Storage Market Snapshot." https://sepapower.org/resource/2019-utility-energy-storage-market-snapshot.

[13] Or four, including the neutral return. See Peterson, D. 2021. "The Importance of Neutral Wire in 3-Phase Systems." *Control Automation*, March 21. https://control.com/technical-articles/the-importance-of-neutral-wire-in-3-phase-systems.

to forecast and therefore difficult to manage. As a result, BTM supply and demand on all three phases could vary in real time. Any imbalances across the phases should be kept in check—left unchecked, imbalances can lead to increases in grid losses. Grid operation and planning complexity also increase as it becomes more difficult for grid operators to manage efficient power system operation. Because power flows in different directions across the three phases, it also directly affects the protection of the distribution system. Variable two-way flows could have a profound impact on relay settings[14] and cause outages to occur. Microgrids may be useful in mitigating such imbalances as they can manage them within their footprint for the larger grid.

A third technological challenge is due to the *shift in generation from large synchronous generators* connected to transmission to increasing numbers of smaller distributed wind and solar generation sources. This shift poses a challenge in the following manner. It is important to maintain alternating current (AC) supplied by the grid at a constant frequency of 60 Hz in order to have a reliable power grid to which millions of appliances that require this constant frequency are connected.[15] Large synchronous fossil-fueled generators help maintain this frequency and restore any deviations that occur because of an imbalance between supply and demand, through the rotating inertia present in these generators. As more generation moves from the transmission side to the distribution side and the type of generation changes, this rotating inertia, which solar and wind generation do not possess, is lost.[16] If not addressed and mitigated, there may be significant frequency oscillations. Solutions can be found through the integration of power electronics,[17] and significant advances are occurring on this front, including the

[14] Relays are devices utilized in a grid to regulate voltages. The latter rise or fall depending on the amount of electricity drawn by each household. Often there can be a large variation and therefore grid operators have to exercise control through these devices to ensure a reliable grid operation. See Horowitz, S.H., A.G. Phadke, and C.F. Henville. 2022. *Power System Relaying*. 5th ed. Hoboken, NJ: John Wiley and Sons; Palmintier, B., R. Broderick, B. Mather, M. Coddington, K. Baker, F. Ding, M. Reno, M. Lave, and A. Bharatkumar. 2016. "On the Path to SunShot: Emerging Issues and Challenges in Integrating Solar with the Distribution System." NREL/TP-5D00-65331. Golden, CO: National Renewable Energy Laboratory. http://www.nrel.gov/docs/fy16osti/65331.pdf.

[15] NERC (North American Electric Reliability Corporation). 2011. *Balancing and Frequency Control: A Technical Document Prepared by the NERC Resources Subcommittee*. Princeton, NJ: NERC. https://www.nerc.com/comm/OC/BAL0031_Supporting_Documents_2017_DL/NERC%20Balancing%20and%20Frequency%20Control%20040520111.pdf.

[16] This challenge is associated with wind and solar technologies whether they are connected to BTM or to higher voltage elements of the delivery system.

[17] Matevosyan, J., B. Badrzadeh, T. Prevost, E. Quitmann, D. Ramasubramanian, H. Urdal, S. Achilles, J. MacDowell, S.H. Huang, V. Vital, and J. O'Sullivan. 2019. "Grid-Forming Inverters: Are They the Key for High Renewable Penetration?" *IEEE Power and Energy Magazine* 17(6):89–98.

development, installation, and maintenance of grid-forming and grid-following inverters[18] for individual applications and in microgrids (see Box 6-2 for further description). Because these investments support the system as a whole, the additional costs for these technologies at the distribution level may be borne by all customers—those with BTM DG and those without—which could have economic, equity, and regulatory implications related to the design of net-metering and its variants.

Increasing Complexity of Operations, Planning, and Forecasting

In addition to technological challenges, the growth of BTM DG increases the complexity of operations, planning, and forecasting. Based on past experience, grid operators have historically held certain expectations about how different types of customers will use electricity from the grid during different times of day, days of the week, and seasons of the year. It is the utility's obligation to ensure that the grid and all interconnected equipment operates safely to prevent injuries to the users. For this purpose, the patterns of electricity usage by customers need to be predictable. As BTM generation and storage increase—in part enabled by net metering and other policies more generally—a number of uncertainties arise with respect to customer demand shapes and the timing and size of any BTM DG exports to the grid. These uncertainties pertain to (1) variable generation and load patterns on both sides of the meter; (2) unpredictable customer behavior related to individuals' habits and personal satisfaction affecting consumption and net BTM DG production (e.g., when these customers run water-heater and air conditioning devices, and operate kitchen appliances, and charge their electric vehicles [see Box 6-3 for a further discussion of these uncertainties]); and (3) random component outages at both sides of the meter. The net variable generation and load at the power distribution level, which would largely stem from BTM DG variabilities, could make the system difficult for the grid operator to manage, particularly under extreme conditions (severe weather and outages due to operator errors, accidents, or intentional physical attacks). Random outages of electrical components at both sides of the meter could also alter BTM DG operation, as any outages could constrain the flow of electricity and affect the system's reliability and resilience. This could have implications for net metering policy

[18] Inverters are important components based on power-electronics and are necessary in converting direct current (DC) power (which comes from PV used in roof-top solars) to AC power. Two broad categories of inverters are grid-following (GFL) and grid-forming (GFM) inverters, with the latter providing more precise control over the AC power. GFM inverters are increasingly needed as more and more large synchronous machines that provide power at transmission get replaced with variable renewables, as they provide more precise control over the delivery of reliable AC power. Song, G., B. Cao, and L. Chang. 2022. "Review of Grid-Forming Inverters in Support of Power System Operation." *Chinese Journal of Electrical Engineering* 8(1):1–15.

BOX 6-2
MICROGRIDS AS A WAY TO INTEGRATE BTM DG

BTM microgrids with enhanced and hierarchical control systems can help to realize the benefits of BTM DG, by integrating various technologies behind the meter of the microgrid itself. These BTM assets can be readily controllable through the microgrid, which can provide various analytics, such as voltage and frequency monitoring. These monitoring analytics could be of significant help in the coordination of BTM DG and other DER with grid operation, peak-shaving, reducing demand uncertainties, and regulation of voltage and frequency, as well as significantly improving grid security and grid-resilience. For example, microgrids can play a role in mitigating the impact of electric power system disturbances under unexpected but catastrophic events like natural disasters and cyber-attacks. Microgrids can be designed so that they can operate in an islanded mode when supply on bulk power systems is disrupted by a natural disaster, functioning as self-contained entities while guaranteeing the continuity of power supplies to local and critically positioned loads.[a] Through the strategic management of on-site resources (including DG and controllable loads), microgrids can further offer ancillary services (e.g., black start[b]) to restore electric power services in parts of the system located outside the microgrid. Many microgrids (e.g., those located in campus settings) may be enabled or financially supported by net metering or virtual net metering.[c] Net metering compensation can reflect this value where microgrids can address some of the technical challenges described above and contribute to strengthening the resilience of electric power system operations.

[a] Wood, E. 2022. "Army Releases Five-Year Microgrid Implementation Plan." Microgrid Knowledge, October 10. https://www.microgridknowledge.com/editors-choice/article/11436859/army-releases-five-year-microgrid-implementation-plan.

[b] A black start is the process of restoring a part of a power grid to operation, after a total or partial shutdown, without relying on the external power transmission network. Jain, H., G. Seo, E. Lockhart, V. Gevorgian, and B. Kroposki. 2020. Blackstart of Power Grids with Inverter-Based Resources. Preprint, NREL/CP-5D00-75327. Golden, CO: National Renewable Energy Laboratory. https://www.nrel.gov/docs/fy20osti/75327.pdf.

[c] Khodayar, M., and M. Shahidehpour. 2013. "Cutting Campus Energy Costs with Hierarchical Control: The Economical and Reliable Operation of a Microgrid." *IEEE Electrification Magazine* 1(1):40–56.

related to the predictability of customer response to the price signals sent by the compensation levels provided.

Over the past several decades, the design of the distribution grid has been dominated by a top-down approach—one in which the utility plans to respond to whatever happens due to customers' withdrawals of electricity from the grid—in terms of the primary and secondary feeders and their interconnections, and the location and type of voltage control devices including transformers, load-tap changers, and capacitor-banks. Through appropriate design and management of these voltage control devices, utilities have operated the power grid and ensured its reliability and power quality.

The growth of BTM DG fundamentally challenges this top-down planning and operations approach, as users may behave in ways operators do not expect. Because users do not exist in a vacuum when it comes to their control and operation of DG, careful coordination between operators and users will be important to ensure appropriate reconciliation between what the users want to do and what the operators need or will allow them to do.

In particular, the impact of BTM on grid operations can be summarized as follows: grid operation will be more challenging because there will be less net load available under the control of grid operators as BTM load increases. A higher BTM load will make it more difficult for grid operators to forecast the net load, as BTM load is quite variable and the BTM load data are mostly unavailable to grid operators. This, in turn, is due to customers exercising their own habits and preferences to charge their cars, discharge their batteries, use their appliances, and participate in programs such as net metering. Complexity could also increase for grid operators with the proliferation of EVs and BTM charging devices.

As a result, updating data collection and analysis, demand forecasting, and other tools are becoming increasingly important to more accurately quantify BTM DG adoption and interactions of customer demand and supply with the local distribution system. Conducting accurate forecasts of BTM DG can help to envisage BTM flexibility requirements, reducing the operational costs of centralized generation and improving the system's efficiency while maintaining its security. To this end, time series forecasting models have gained prominence.[19] Grid operators have leveraged forecasting models for estimating BTM load, generation, and market prices, and for estimating time-horizons for predicting BTM load and generation for quantifying risks and reducing uncertainties to lead to better planning operational strategies.

TECHNOLOGY INTERSECTION 2: EMERGING ADVANCES IN POWER ELECTRONICS, STORAGE, COMMUNICATIONS, AND CONTROLS

The seamless integration of significant amounts of DG and flexible loads and exports necessitates advances in technologies at multiple levels to ensure reliable grid operation. Some technologies can be specifically encouraged or addressed directly as part of net metering compensation structures and levels so that, for example, they encourage types, locations, and timelines for DG adoption and operation that enhance or do not

[19] Time series models account for internal structural relationships within data, often employed for the purposes of forecasting future outcomes. See NIST. 2022. *Engineering Statistics Handbook: 6.4. Introduction to Time Series Analysis*. Washington, DC: NIST. https://www.itl.nist.gov/div898/handbook/pmc/section4/pmc4.htm.

BOX 6-3
ELECTRIC VEHICLES, CHANGING LOAD PATTERNS, AND NET METERING

One of the significant challenges that BTM DG presents is a change in net load patterns. As electric vehicle (EV) charging increases, load patterns may change significantly; peak loads may begin to shift from daytime to evenings, when most EVs would be charged. The locations and capacities of charging stations, driving patterns, and demographics can all have an impact on these load-shifts. With such a load-shift, many of the key components in a distribution grid may be stressed. This in turn changes the load patterns in a distribution grid posing direct challenges to the grid operator for both planning and operations. There are several indirect challenges that should be noted as well. One of them pertains to the health of transformers, key devices in the grid network that ensure reliability. As EVs proliferate, and vehicle-to-grid interactions increase, standard load patterns with higher usage during the day and minimal usage during the night can change drastically. Current transformers are designed for such a load pattern. Increased usage during the night due to increased EV charging can place additional load on transformers by not permitting the transformer oils to cool, thereby directly affecting grid reliability. Excess availability of solar PV generation in midday could negatively affect grid operations, as net load (i.e., total load minus solar PV generation) could drop below the minimum thermal generation capacity, causing thermal generators to shut down automatically. Midday pricing could be implemented to allow additional BTM generation to be made available to the public, and EVs could be incentivized to use this excess capacity to prevent thermal unit shutdowns. As discussed in Chapter 4, net metering compensation should align with the value or costs to the system BTM DG provides. In this instance, where it can be encouraged to provide power for EV charging and relieve transformer loads.

adversely affect the reliability of the grid. Whether or not net metering compensation approaches require such technologies, the distribution utility may need to add them to assure reliable operations of the grid.

The first technology relates to power electronics that allow the transformation from DC to AC through the use of *inverters*. "Smart" inverters that operate autonomously, both in grid-following and grid-forming modes, and which coordinate with each other and the grid, are being required in certain jurisdictions for solar (and other BTM DG) to be eligible for net metering.[20]

Storage is another technology and, as discussed earlier, perhaps the most attractive option for addressing the variability and uncertainties associated with increasing DG and DER penetration. Advanced batteries, thermal storage devices, and emerging

[20] CPUC. 2021c. "Rule 21 Interconnection." https://www.cpuc.ca.gov/rule21.

technologies based on hydrogen are all options that are seeing exponential advances.[21] The rapid growth of electric vehicle adoption provides another opportunity for potentially cost-effective storage that could be integrated into grid services and/or supply if technically enabled to do so. Advances are being reported, both in V1G (also known as managed charging and/or smart charging) and vehicle-to-grid (V2G) technologies, where the former consists of uni-directional control of electric vehicles that switches off charging on demand and the latter consists of bi-directional control of the vehicles by switching between a charging and a discharging mode.[22] With a concerted push by decarbonization advocates, policymakers, and stakeholder groups toward Electrify Everything,[23] there is a significant opportunity to make use of these mobile storage assets for grid reliability, security, and resilience.

While such technological advances support the availability of resources for ensuring power balance, and therefore grid reliability, utilities and grid operators may only be able to leverage them if there exists a strong *communications* backbone and a SCADA system[24] to enable the coordination needed between these critical assets and control rooms located at appropriate locations in the distribution grid. The increasing growth of DER, accompanied by growth in advanced metering infrastructure (AMI), automation

---

[21] Mongird, K., V. Viswanathan, J. Alam, C. Vartanian, V. Sprenkle, and R. Baxter. 2020. *2020 Grid Energy Storage Technology Cost and Performance Assessment*. Publication No. DOE/PA-0204. Richland, WA: Pacific Northwest National Laboratory, and Somerville, MA: Mustang Prairie Energy. https://www.pnnl.gov/sites/default/files/media/file/Final%20-%20ESGC%20Cost%20Performance%20Report%2012-11-2020.pdf.

[22] Markiewicz, D., B. Wechtenhiser, J. Brendlinger, M. Campbell, M. Tims, and J. Yanuck. 2018. "Vehicle-to-Grid Testing and Demonstration Final Report." CEC-500-2018-024. Prepared for California Energy Commission by Concurrent Technologies Corporation. https://www.energy.ca.gov/sites/default/files/2021-06/CEC-500-2018-024.pdf.

[23] Regeneration. 2022. "Electrify Everything." https://regeneration.org/nexus/electrify-everything; Roberts, D. 2017. "The Key to Tackling Climate Change: Electrify Everything." *Vox*, October 27. https://www.vox.com/2016/9/19/12938086/electrify-everything.

[24] SCADA stands for Supervisory Control and Data Acquisition. It corresponds to information system architecture that consists of computers, networked data communications, and graphical user interfaces so that it can receive real time data on power system processes and related equipment, which are in turn used by grid operators and supervisors to control, mitigate contingencies, and optimize power system operations. Barnes, K., B. Johnson, and R. Nickelson. 2004. *Review of Supervisory Control and Data Acquisition (SCADA) Systems*. Idaho Falls, ID: Idaho National Engineering and Environmental Laboratory. https://inldigitallibrary.inl.gov/sites/sti/sti/3310858.pdf.

devices such as reclosers, controllable capacitor banks, and voltage regulators, and smart inverters, necessitate the deployment of advanced communication technologies.[25,26]

In addition, a *distributed decision-making architecture* is needed to manage power injections, load reductions, and storage in this distributed environment. Utilities and grid operators have to make decisions with significant and varying uncertainties in the knowledge of the system state (due to latencies in communications or faults and outages in equipment). Moreover, as DG grows in scale, whether behind the meter or otherwise, the amount of data grows exponentially, leading to a need for data management at scales to which utilities are not accustomed. There are significant advances being made in the requisite distributed control technologies that may allow robust and accurate decisions at the appropriate level of granularity and across all locations, and the opportunity for reliable and resilient integration of DER, including DG, across the grid. However, such technical solutions must be implemented in conjunction with (or even in anticipation of) rapid deployment of DG for local systems to perform reliably over time.

These advances introduce new challenges as well. First, *cybersecurity* poses a real and growing challenge, as discussed further below. Another issue is *control* of BTM DG with disparate ownership. Net metering policies can be shaped to encourage customer flexibility with respect to DG, as well as DER, operation, and consumption in general. Policies can also be developed with opportunities for demand response (e.g., direct load control) and transactive energy in mind.[27] For example, an internal price (through net

[25] Over the past two decades, communication technologies have evolved at a rapid pace, with new generations of technologies continually evolving. Notable ones are wireless technologies which have evolved through four generations leading to current exploration of 5G networks, and discussions of 6G not far behind. Internet of Things (IoT)-enabled devices have provided a new platform for grid operators to facilitate flexible consumption. A compelling demonstration is the synchronous reduction of load by 700 MW through an IoT-coordinated control of a large number ofsmart thermostats during a total solar eclipse in 2017. It should be noted that this occurred during a solar eclipse. See Tsao, S. 2017. "PJM, CAISO Detail Load Changes During Solar Eclipse." *S&P Global Market Intelligence*, September 13. https://www.spglobal.com/marketintelligence/en/news-insights/trending/h-gQiWXabDmfyAvghazbsw2.

[26] Zhou, Q., M. Shahidehpour, A. Paaso, S. Bahramirad, A. Alabdulwahab, and A. Abusorrah. 2020. "Distributed Control and Communication Strategies in Networked Microgrids." *IEEE Communications Surveys and Tutorials* 22(4):2586–2633.

[27] Hammerstrom, D.J., R. Ambrosio, T.A. Carlon, J.G. DeSteese, G.R. Horst, R. Kajfasz, L.L. Kiesling, et al. 2007. *Pacific Northwest GridWise Testbed Demonstration Projects: Part I. Olympic Peninsula Project*. Richland, WA: Pacific Northwest National Laboratory. https://www.pnnl.gov/main/publications/external/technical_reports/PNNL-17167.pdf.

metering compensation) can be used as a control signal to consumers to adjust their consumption.[28]

Several transactive energy studies have been published in the past few years[29,30] identifying advantageous grid-wise benefits such as reduced demand volatility, reduction of peak demand, lower electricity prices, and improved capacity utilization. Transactive energy systems create revenue opportunities for residential DG and DER owners, and do so in a way that can more accurately reflect the true value of DG and DER operation at that time, and as it changes over time. Transactive energy systems also provide a platform to incorporate equity considerations, for example, by coupling a transactive offering with targeted or means-tested financing and education for technology purchases. A NARUC report,[31] discussing electric system digitalization talks about transactive energy and other approaches that can serve similar objectives as those underlying net metering. Transactive energy is also adaptable to many use cases, from a utility using it to help it achieve its operational objectives, to a microgrid operator or building operator using it to manage and balance production and consumption, to a retail energy provider using it to enable its risk management across its customer portfolio, and in light of its wholesale long-term contracts and prices in wholesale power markets. Significantly more studies need to be carried out to understand issues related to tariff, interconnection, and ownership considerations related to who manages or who has technical control over

---

[28] Demand response programs offer models that may be applied to BTM distributed generation through NEM compensation design. A typical illustration of this project demonstrated peak load reduction and energy cost saving through the coordination of multiple water pumps, diesel generators, and heating loads of several residential houses. Kangping L., F. Wang, Z. Mi, M. Fotuhi-Firuzabad, N. Duić, and T. Wang. 2019. "Capacity and Output Power Estimation Approach of Individual Behind-the-Meter Distributed Photovoltaic System for Demand Response Baseline Estimation." *Applied Energy 253*:113595.

[29] Cazalet, E.G, M. Kohanim, and O. Hasidim. 2020. *Complete and Low-Cost Retail Automated Transactive Energy System (RATES)*. CEC 500-2020-038. Prepared for the California Energy Commission. Encino, CA: Universal Devices, Inc. https://www.energy.ca.gov/sites/default/files/2021-05/CEC-500-2020-038.pdf. Appendix B provides information on the participants and the nature of their uses of the technologies and degree of active participation.

[30] PNNL (Pacific Northwest National Laboratory). *The Distribution System Operator with Transactive (DSO+T) Study: Executive* Summary. Richland, WA: PNNL. https://www.pnnl.gov/sites/default/files/media/file/EED_1574_BROCH_DSOT-ExecSumm_v11.pdf.

[31] Kiesling, L. 2022. *Digitalization in Electric Power System and Regulation: A Primer*. Knowledge Problem LLC. https://pubs.naruc.org/pub/17AB6931-1866-DAAC-99FB-8D696CC1944A.

such resources under different circumstances,[32] in order to evaluate the scalability of transactive energy to help realize the full benefit of DG for the customer and the grid.

The above discussion illustrates that many of the engineering challenges of increasing BTM DER can be addressed through the deployment of appropriate advanced technologies. Customer deployment of some of these technologies related to DG could be encouraged through net metering compensation.

TECHNOLOGY INTERSECTION 3: CYBERSECURITY AND RESILIENCE

Cybersecurity and Privacy Considerations

As the number of cyber-attacks aimed at disrupting electric power supplies increases globally,[33] electricity system operators face a new frontier for managing resilience.[34] Cyber-attacks can further magnify the impacts of physical disruptions. While cyber-secure BTM DG systems could offer a significant potential to sustain power grid services under extreme operating conditions, particularly those with highly reliable and resilient power supplies, they too are also subject to cyber incidents. Moreover, cyber vulnerabilities in

[32] In its September 2020 "Order 2222," The Federal Energy Regulatory Commission (FERC) enabled distributed energy resources to compete in all regional wholesale electric markets. When fully implemented, DER will participate in wholesale markets, providing "capacity, energy, and ancillary services alongside traditional resources." FERC explains, "[m]ultiple DERs can aggregate to satisfy minimum size and performance requirements that they might not meet individually." FERC states Order 2222 is intended to "empower new technologies to come online and participate in wholesale markets on a level playing field to further enhance competition, encourage innovation and drive down costs for consumers." In June 2022 FERC issued a first order regarding compliance with Order 2222. See FERC. 2020b. "FERC Opens Wholesale Markets to Distributed Resources: Landmark Action Breaks Down Barriers to Emerging Technologies, Boosts Competition." September 17. https://www.ferc.gov/news-events/news/ferc-opens-wholesale-markets-distributed-resources-landmark-action-breaks-down; and FERC. 2022. "FERC Acts on First of Order No. 2222 Compliance Filings." June 16. https://ferc.gov/news-events/news/ferc-acts-first-order-no-2222-compliance-filings.

[33] For instance, cyber attackers switched off breakers remotely in the Ukrainian electric power system with the help of a malware named BlackEnergy malware on December 23, 2015, which left approximately 30 substations disconnected and approximately 225,000 customers without power for several hours. Cybersecurity & Infrastructure Security Agency. 2021. "Cyber-Attack Against Ukrainian Critical Infrastructure." https://www.cisa.gov/news-events/ics-alerts/ir-alert-h-16-056-01; E-ISAC (Electricity Information Share and Analysis Center). 2016. *Analysis of the Cyber Attack on the Ukrainian Power Grid: Defense Use Case*. Washington, DC: E-ISAC.

[34] Pp. 4, 9, and 56–60 in NASEM. 2021. *The Future of Electric Power in the United States.* Washington, DC: The National Academies Press. https://doi.org/10.17226/25968. See also NASEM. 2017. *Enhancing the Resilience of the Nation's Electricity System*. Washington, DC: The National Academies Press. https://doi.org/10.17226/24836.

BTM DG operations can potentially trigger a widespread outage in electric power systems. It is thus important to identify these vulnerabilities and deploy effective measures to address cyber threats.

The focus of BTM DG cyber security is to retain the quality and continuity of supplied power. BTM device malfunction may occur when the device is reconfigured and manipulated in an unauthorized manner. If a BTM device is compromised, attackers can leverage control over the device and hinder normal functions of the associated sensors and/or actuators (e.g., altering the data that will be reported to the grid control center, and denying the implementation of supervisory control commands). In addition to operators' unintended errors, software applications are potentially corrupted and modified by viruses hidden in the BTM DER servers' operating systems. Hence, it is critical to strengthen BTM DER cyber security capabilities for sustaining DER (including DG) services in the event of cyber incidents to protect the grid and BTM DER customers.

As more and more devices are connected to the grid, they pose correspondingly increasing cyber risks that need to be addressed by local grid operators. Failure to address such considerations will weaken the reliability of the grid, potentially offsetting to some degree the other grid and customer-specific resilience benefits of deployment of DG. Overcoming the cybersecurity risks and ensuring a reliable operation of the grid will entail investments in secure communication and control technologies.

Increased penetration of rooftop solar also implies increased communication, which implies that personal information that can identify an individual directly or indirectly may be revealed, that is, privacy may be compromised. As increased communication is essential to ensure grid reliability, particular attention has to be paid to ensure that sensitive lifestyle information is protected regarding when the home is occupied or when the occupants are awake, or how and when they use various appliances. Related information about billing history, account information, and DER information has to be protected at various levels of security. Fraud protection has to be ensured; this includes energy consumption that is attributed to a different location, and also that any customer data collected must be kept secure and not revealed to any third party.[35]

[35] NIST (National Institute of Standards and Technology). 2010. *Guidelines for Smart Grid Cyber Security*, pp. 1–3. NISTIR-7628. Washington, DC: NIST. https://nvlpubs.nist.gov/nistpubs/ir/2010/NIST.IR.7628.pdf.

Resilience Considerations

As the amount of capacity and installations of BTM DG grow, many utilities and the federal government are addressing how the grid can remain resilient,[36] not only in the face of cyber-attacks but also related to extreme weather (tornados, hurricanes, winter storms, etc.). Depending on how DG installations are integrated into the grid, the electric system may become more resilient to widespread power outages as consumers produce and have better access to on-site power. If DG is configured so that it can provide resilience, not only to the consumer but also to the distribution grid, the corresponding net metering compensation should reflect this value.

Various DG technologies can be utilized to provide resilience, the most common of which is backup generation. There are however more complex scenarios, which include a combination of DG and storage.[37] For example, DG, backup generation, storage, or reduction of demand need not be located at the same house or building. A well-thought-out design in a supporting cyber-layer of smart digital devices that monitor, process, communicate, and alert other devices (e.g., in a different location, in the form of a microgrid, or a community solar device, or even rooftop solar at a neighboring house) that are connected to the overall network can allow all of these technologies to step in, alleviate peak loads, supply critical nodes such as hospitals and fire stations, and help with the overall grid resilience.[38] Policies and regulations may need to be modernized to allow the most advanced applications to enable the highest level of resilience.[39]

[36] DOE Office of Electricity. 2019. "North American Energy Resilience Model." https://www.energy.gov/sites/prod/files/2019/07/f65/NAERM_Report_public_version_072219_508.pdf. This report defines a resilient grid as one that has the ability to anticipate, withstand, and recover critical loss-of-supply resulting from low-probability, high-impact threats. NASEM. 2017. *Enhancing the Resilience of the Nation's Electricity System*. Washington, DC: The National Academies Press. https://doi.org/10.17226/24836.

[37] For example, in northern regions during the winter, heavy snows and low solar illumination can make providing a household's electric power solely with a solar + storage system cost prohibitive unless other backup generation is installed. Currently, for short term power outages, the least capital-intensive solution is to provide backup to the DG + storage with a fossil-fired generator. (Generators are widely commercially available that can power either key loads in a building [e.g., refrigerator] or the entire building from several sources of fuel [e.g., diesel, propane or natural gas].) However, a generator may not be economic for extended use because of the relatively high cost of both the generator and fossil fuels.

[38] Grid-level resilience corresponds to a basic restoration of electricity over a large area (such as a street, a township, or a district), whereas local resilience corresponds to restoration of electricity within a building such as a hospital using alternate electrical supply. See NASEM. 2017. *Enhancing the Resilience of the Nation's Electricity System*. Washington, DC: The National Academies Press. https://doi.org/10.17226/24836.

[39] See NASEM. 2017. *Enhancing the Resilience of the Nation's Electricity System*. Washington, DC: The National Academies Press. https://doi.org/10.17226/24836.

SUMMARY

Public policy, including net metering, supporting distributed renewable generation, declining costs in solar and storage, and exponential advances in technologies related to metering, communication, computation, and power electronics, has led to a steady increase in DER in general and DG in particular. This chapter outlined the challenges in the integration of BTM DG and other DER, the technological advances that provide tools for effectively addressing these challenges, and various ways in which these technologies can support the grid, including improving reliability and resilience while ensuring distributed renewable integration.

With advances in various DER technologies, it can be expected that DG costs due to solar PV installation and associated storage will continue to decline, making it more attractive for consumers to adopt renewable generation, especially rooftop solar. Correspondingly, DG adoption with net metering compensation can be expected to increase. To simultaneously address the accompanying challenges of variable demand, grid reliability, and cyber-physical security, investment in technology on the distribution system is essential to, among other things, provide visibility to grid operators so that they can assure the stable operation of the system while also unlocking the potential of DG and DER to provide grid services and value to the system.[40] Without investment in such grid technologies, not only might the potential for BTM DG to provide grid services be stymied, but also deeper deployment might have negative consequences on grid reliability. Policymakers and regulators need to understand the importance of such grid investments in conjunction with increasing deployment of DG as they consider the framework for allowing utilities to recover these costs.

FINDINGS

Finding 6-1: Physical constraints that govern electricity flow introduce several challenges as BTM DG increases: reverse flow of electricity, significant uncertainties in load, and variations in load. These in turn can lead to voltage fluctuations and stress grid equipment, causing outages. As DER penetration, especially that of DG, increases, the need for investments in the distribution grid in order to ensure reliability and resilience becomes imperative.

[40] Net metering rules and rate structures can also include, encourage, or require transparency to grid operators, which will enhance the value of BTM DG and enable grid operators to address any potential negative consequences for the electricity system.

Finding 6-2: In addition to direct investments to integrate BTM DG and other DER, investments in communications, visibility, security, and digital power electronics are essential to capture the value of DER and DG for all customers and the system as a whole.

Finding 6-3: Investments in technologies to integrate BTM DG and other DER can lead to a reliable and resilient grid-edge, easing the burden on future investments in distribution and transmission expansion.

RECOMMENDATIONS

Recommendation 6-1: There need to be direct investments in the distribution system to integrate increasing amounts of BTM DG such as rooftop solar, as well as other DER, including smart buildings management systems, electric vehicles, and charging infrastructure, to ensure the continued safe and reliable operation of the grid and provision of grid services. Investments will also be required to improve grid visibility to suitably site and operate BTM DG, as well as provide efficient price signals, such that the DG can provide system benefits, particularly local and grid resilience when normal service is disrupted. These investments in the distribution grid have to occur simultaneously with BTM DG deployment.

Recommendation 6-2: In order to make the best use of DG and DER, utilities must make investments to integrate, increase the visibility of, manage (either directly or indirectly through price signals), and reduce barriers to customer and DG provider management of these technologies. With utility or non-utility control or intelligent management, DG can provide more value. The corresponding compensation through net metering and its variants could reflect this higher value. Overall, democratization of the grid, with increased public and private partnerships, can be very valuable.

Recommendation 6-3: Investments in distribution system technologies toward integration of DG and DER have to be accompanied by revisions in policies and state and federal utility regulations to facilitate cost recovery of these investments.

7

Regulatory, Legal, and Market Considerations

Prior chapters have focused on several important considerations in the design of net metering and variant ratemaking approaches to meet a range of short- and long-term objectives of policymakers, consumers, and others. Chapter 3, for example, explains the trends in adoption of net metering and in state regulators' consideration of whether to modify net metering design over time and, if so, how. Chapter 6 described technical issues associated with the grid's ability to function reliably with the injection of power from behind the meter distributed generation (BTM DG) and the need for investment in the distribution system to capture the benefits it can potentially provide. Chapter 4 discussed the importance of economic efficiency principles in designing utility rates generally and in establishing levels of compensation for BTM resources' exports to the grid. Chapter 5 addressed key equity considerations that need to be taken into account in designing rates for BTM resources while assuring the affordability of basic electricity services and access to DG benefits for all, including those not able to adopt such BTM resources themselves.

In practice, utility ratemaking takes technology, efficiency, and equity issues into account while also balancing other and sometimes competing goals.[1] Indeed, legislatures and regulators make policy decisions about electric utility services and rates, including net metering approaches, that often rely on considerations other than efficiency, equity, and technical issues in setting rates.

This chapter examines the range of legal and regulatory considerations that, among other things, affect the pace and character of adoption and evolution of net metering policies across states and across utility service territories even within a state. The focus here is on understanding these regulatory and legal topics from the vantage point of their interaction with net metering issues, and to help identify policy considerations that can enable technology deployment and competitive markets for innovative

[1] Solar Electric Power Association. 2013. "Ratemaking, Solar Value and Solar Net Energy Metering—A Primer," pp. 5–6. Washington, DC: Solar Electric Power Association. https://www.energy.gov/sites/prod/files/2015/03/f20/sepa-nem-report-0713-print.pdf.

technologies, and that support a decarbonizing, equitable, economically efficient, and resilient electricity system.

This discussion begins with an explanation of federal versus state regulatory jurisdiction over the electricity industry and BTM resources, and then explores several regulatory, legal, and market considerations relating to net metering, largely at the state level. This chapter includes:

- The jurisdictional split over the regulation of wholesale versus retail sales of electricity.
- Status regarding federal regulation of net metering.
- State regulation and net metering policy, including traditional electric utility ratemaking principles, power acquisition principles, and other policies pertaining to the evolving relationships between electricity customers, their utility service providers, and non-utility electricity product and service providers.
- The electric utility's traditional obligation to serve and its implications for the provision of utility service to customers that adopt technologies that provide part of their power needs.[2]
- Issues related to rights and responsibilities of different entities (e.g., the owner of on-site equipment that feeds supply into the grid; the utility that interconnects BTM resources onto the system; and regulators that set net metering rates) involved in DG that feeds into the grid, given different contractual or tariffed services.
- Insights about how net metering interacts with local circumstances, including the penetration levels of BTM resources on a local system, different electric utility ownership forms, electric industry structure, the presence or absence of different regulatory policies (e.g., revenue decoupling), and the existence of complementary federal and state policies affecting the adoption of BTM resources.
- Principles for the constructive design of policies that aim to launch and invigorate markets for the supply and demand for new technologies, including clean DG.[3]
- Findings and recommendations.

[2] The committee's statement of task includes "The committee will consider how new technologies and their attendant new capabilities alter the roles of the participants in the system, the requirements of each participant to fulfill those roles, and the challenges and opportunities present."

[3] The committee's statement of task includes "The study will give recommendations on key principles for policymakers when considering net metering and alternative policies."

THE JURISDICTIONAL SPLIT OVER REGULATION OF WHOLESALE VERSUS RETAIL SALES OF ELECTRICITY

In the United States, regulation of the electricity industry is split between federal, state, and local governments, and by activity:[4]

- *Retail sales:* State and local regulation and control. The prices, terms, and conditions of sales to end-use electricity customers are subject to state regulation for services provided by investor-owned electric utilities and to the decisions of elected or appointed boards of directors of publicly owned electric utilities.[5]
- *Wholesale sales:* Federal regulation. The Federal Power Act authorizes and directs the Federal Energy Regulatory Commission (FERC) to regulate, among other things, the prices, terms, and conditions of sales of electricity (i.e., sales for resale; wholesale sales) and transmission service in interstate commerce.[6]

This means that the design of rates for the *purchase of electricity by customers with BTM resources* is subject to pricing decisions of state regulators for investor-owned utilities and of boards of publicly owned distribution utilities. *Exports of power for sale or for compensation* (e.g., to the utility or an independent power supplier or other entity) tend to be subject to state and local regulation and decisions rather than federal, except potentially (as discussed further in this chapter) where such exports take on the characteristics that create conditions resembling a relatively persistent export of electricity to the grid—and thus, a sale for resale.

[4] For a detailed discussion of federal versus state regulatory jurisdiction and policy in the electric industry, see NASEM (National Academies of Sciences, Engineering, and Medicine). 2021. Chapter 3 in *The Future of Electric Power in the United States*. Washington, DC: The National Academies Press. https://doi.org/10.17226/25968 (hereafter NASEM, *The Future of Electric Power*).

[5] Here, "publicly owned utility" refers to utilities that are not owned by investors. Publicly owned utilities thus include municipal utilities, cooperative utilities, and utilities that are public special purpose districts. There are a few states (e.g., Vermont 30 V.S.A. § 203 Jurisdiction of Certain Public Utilities) where the state regulatory agency has some jurisdiction of publicly owned utilities.

[6] 16 U.S.C., 824(a). "The Federal Power Act (FPA) authorizes the Federal Energy Regulatory Commission (FERC) to regulate 'the sale of electric energy at wholesale in interstate commerce,' including both wholesale electricity rates and any rule or practice 'affecting' such rates." 16 U.S.C. §§ 824(b), 824d(a), 824e(a). But it places beyond FERC's power, leaving to the states alone, the regulation of "any other sale" (i.e., any retail sale) of electricity. § 824(b). FERC (Federal Energy Regulatory Commission). 2016. *Federal Energy Regulatory Commission v. Electric Power Supply Association, et al.* U.S. Supreme Court, No. 14-840. https://www.justice.gov/sites/default/files/osg/briefs/2015/01/28/ferc_v._epsa_app.pdf (hereafter *FERC v. EPSA*). Note that because most of Texas (i.e., the part known as the Electric Reliability Council of Texas [ERCOT]) is electrically isolated from other states' electricity systems, wholesale electricity in Texas does not trigger the "in-interstate-commerce" provision of the Federal Power Act, and FERC does not have jurisdiction over the terms and conditions of sales for resale in the ERCOT portions of Texas.

FEDERAL RATE REGULATION AND NET METERING

To date, relatively few, narrow circumstances exist where FERC may—and does—regulate rates for exports into the grid from BTM resources. For example, where distributed energy resources (DER) (e.g., demand response; rooftop solar) are offered as supplies in organized wholesale competitive markets,[7] prices (and, thus, compensation for exports) in those markets are FERC-regulated and established through market-based mechanisms, even though the resources can be located on and interconnected with the distribution system.[8]

FERC regulates the rates of BTM generation as "qualifying facilities" (QFs) under the Public Utility Regulatory Policies Act (PURPA)[9] where that generating capacity (e.g., rooftop solar or combined heat and power cogeneration facilities) is sized larger than the host end-use-customer's loads and thus causes that electricity customer to be a chronic, net-exporter to the grid.[10] Under PURPA, QFs must be paid a price set at the avoided cost of the purchasing utility.[11] The PURPA avoided cost rate is determined by the state public utility regulatory authority.[12]

That said, a customer seeking to add on-site solar and storage equipment that exceeds the on-site loads for some period of time—say, 2 years ahead of purchasing an electric vehicle which would then increase on-site electricity demand—could unwittingly trigger

---

[7] "Organized markets" is the term used to describe wholesale power markets established in and administered by Regional Transmission Organizations (RTOs) and Independent System Operators (ISOs). All of the nation's RTOs operate mandatory centralized, bid-based, security-constrained wholesale electric energy markets, while a few RTOs (i.e., New England ISO-NE, New York Independent System Operator NYISO, and PJM Interconnection) also operate centralized wholesale electric capacity markets. FERC. 2021. "Electric Power Markets." https://www.ferc.gov/electric-power-markets.

[8] *FERC v. EPSA*; Rossi, J. 2016. "Federalism and the Net Metering Alternative." *The Electricity Journal* 29:13–18 (hereafter Rossi 2016). Note that FERC Order 2222 requires RTOs and ISOs to allow entities that pool or aggregate capabilities from multiple DER to offer those resources directly into the wholesale markets as capacity, energy, and/or ancillary services. This change may make it more attractive for DER and DG aggregation as a viable business, as compared to relying on net metering alone. Zhou, Q., M. Shahidehpour, A. Paaso, S. Bahramirad, A. Alabdulwahab, and A. Abusorrah. 2020. "Distributed Control and Communication Strategies in Networked Microgrids." *IEEE Communications Surveys and Tutorials* 22(4):2586–2633.

[9] For additional information on Public Utility Regulatory Policies Act qualifying facilities, see https://www.ferc.gov/qf.

[10] The federal 2005 Energy Policy Act "catalyzed distributed generation under net metering ... by encouraging state adoption of net metering policies." Revesz, R.L., and B. Unel. 2016. *Managing the Future of the Electricity Grid: Distributed Generation and Net Metering*. Public Law Research Paper No. 16-09. New York: New York University School of Law. https://policyintegrity.org/files/publications/Managing_the_Future_of_the_Electricity_Grid.pdf (hereafter Revesz and Unel 2016).

[11] Revesz and Unel 2016.

[12] NRRI (National Regulatory Research Institute). 2022a. "PURPA Tracker." https://www.naruc.org/nrri/nrri-activities/purpa-tracker.

becoming a FERC-regulated QF under PURPA during that interim period, with potential tax consequences associated with the production and sale of power to the local utility and with different compensation consequences for the power exported to the grid (if the net-metering and net billing compensation differs from the posted avoided cost PURPA rate offered by the utility). Currently this is more of a theoretical possibility rather than something that happens in practice, but it could become more of an issue, depending on the relationship(s) between on-site use and total BTM outputs (and net exports).

To date, FERC has otherwise declined to regulate rooftop solar and some other BTM DER[13] and has taken what one legal expert has characterized as a "hands-off approach to state net metering."[14] FERC's hands-off policy was tested as recently as 2020 when a group asked FERC to exercise its jurisdiction over exports from BTM resources, but the agency declined to deviate from prior positions.[15] Given the intensity of interest in the question of FERC jurisdiction over net metering, it would not be surprising to see this matter raised again in the future, either at FERC itself or in the courts. That said, given the current state of play on these federal and state jurisdictional issues, the committee has focused attention on actions by states and electric distribution utilities.

[13] "FERC decisions consistently disclaim federal jurisdiction over net metering credits, based on the principle that there is no wholesale sale of energy unless a customer consistently produces sufficient excess energy over the netting period to become a net seller rather than purchaser. Rossi 2016, p. 14. See also Hempling, S., C. Elefant, K. Cory, and K. Porter. 2010. *Renewable Energy Prices in State-Level Feed-In Tariffs: Federal Law Constraints and Possible Solutions*. Technical Report NREL/TP 6A2 47408. Golden, CO: National Renewable Energy Laboratory. https://www.nrel.gov/docs/fy10osti/47408.pdf.

[14] Ari Peskoe at the Harvard Law School has stated that the "bottom line is that in 2001 FERC determined it had no jurisdiction over energy transfers between a retail customer and her utility so long as the consumption was greater than production" over some state-determined period. They reiterated it as recently as 2018 and 2019 in its landmark storage orders that require RTOs to provide participation models for energy storage," pp. 2–3. Peskoe, A. 2020. "Ari Peskoe, Director of the Electricity Law Initiative, on Comments Filed in Opposition to Net Metering Petition, June 10, 2020." Environmental and Energy Law Program, Harvard Law School. http://eelp.law.harvard.edu/wp-content/uploads/Transcript-of-Ari-Pesko-on-FERC-filing-6-10-20.pdf.

[15] Raskin, D.B., and R.L. Roberts. 2020. "Petition for Declaratory Order of New England Ratepayers Association Concerning Unlawful Pricing of Certain Wholesale Sales." United States of America before the Federal Energy Regulatory Commission. New England Ratepayers Association, Docket No. EL20-42-000. https://s3.documentcloud.org/documents/6843679/Nera-Proposal.pdf; TFERC dismissed the petition on procedural grounds; FERC. 2020c. *Order Dismissing Petition for Declaratory Order*. 172 FERC 61,042. New England Ratepayers Association, Docket No. EL20-42-000. https://www.ferc.gov/sites/default/files/2020-08/EL20-42-Order.pdf.

STATE RATE REGULATION, SUPPLY ACQUISITION, AND NET METERING

Most of the action on net metering policy thus resides at the state and local level—at state legislatures and/or before state utility commissions, or at the boards of publicly owned utilities that set policy directions, rates, and resource procurements of retail electric utilities.

State Action on Net Metering

States' net metering policies—whether driven by legislative or regulatory actions—ultimately stem from the statutory authorities in a state. As described in Chapter 3, a handful of states started to experiment with adopting net energy metering requirements in the early 1980s. By the end of the 1990s, net metering programs were offered to some or all electricity customers in 22 states.[16] By that point, 6 states had passed laws explicitly addressing net metering while another 14 states adopted net metering by regulation and thus through their pre-existing statutory authorities.[17] As of 2021, 60 percent of the states and the District of Columbia had net metering policies or successor tariffs in place (see Chapter 3). Most of the remaining states had policies requiring compensation for distributed generation exports to the grid.

Clearly, decision makers in those local venues have to make ratemaking decisions—about net metering and about the design of other rates—that attempt to balance a number of objectives relating to utility service. As the 2021 National Academies' study on *The Future of Electric Power in the United States* observed,

> In many respects, local utilities bear the front-line responsibility to meet the public's and policy makers' many and varied expectations for a safe and reliable electric system—(1) affordability and equity, (2) reliability and resilience, and (3) sustainability and clean-power generation....
>
> State regulators and boards of publicly owned utilities tend to have to make decisions that take into consideration many sometimes competing goals:

---

[16] Wan, Y.H., and H.J. Green. 1998. "Current Experience with Net Metering Programs." CP-500-24527. Golden, CO: National Renewable Energy Laboratory. Paper presented at WINDPOWER '98, Bakersfield, CA. https://www.nrel.gov/docs/legosti/old/24527.pdf.

[17] Across the states, where a state statute requires net metering, this requirement tends to apply to all utility service territories in that state; where regulatory agencies require net metering, the policy typically only applies to investor-owned utilities in the state. See Wan, Y.H., and H.J. Green. 1998. "Current Experience with Net Metering Programs." CP-500-24527. Golden, CO: National Renewable Energy Laboratory. Paper presented at WINDPOWER '98, Bakersfield, CA. https://www.nrel.gov/docs/legosti/old/24527.pdf.

- Ensuring that all households have access to basic, affordable electricity.
- Protecting the core regulatory compact[18] that underlies the monopoly franchise...while also allowing for and creating incentives for customer choice and innovation in business models and technological adoption....
- Respecting precedent in law and regulation, while also allowing for changes that accommodate the needs of and conditions on the changing grid and society....
- Proceeding, in many states, with actions to reduce greenhouse gas (GHG) emissions from the electric system on an urgent timeline, while also attending to potential stranded costs.[19]

This ongoing need to balance various considerations sets the context in which state policymakers make decisions about net metering.

Principles of Regulation of Retail Rates and Net Metering

Regulatory decisions with regard to net metering tariff design also need to fit within the larger context in which utility rates are set and approved by regulators (for investor-owned utilities) and by boards of publicly owned and member-owned distribution utilities.

With regard to overall ratemaking policy, over the last century of electric utility regulation in the states, regulators have generally adhered to a number of principles as they review utilities' rate proposals. Traditional utility ratemaking principles were explained decades ago in the seminal book on utility regulation by James Bonbright and colleagues,[20] and

[18] Jim Lazar of the Regulatory Assistance Project describes the regulatory compact like this: "Effectively, regulation constitutes an agreement between a utility and the government: the utility accepts an obligation to serve in return for the government's promise to approve and allow rates that will compensate the utility fully for the costs it incurs to meet that obligation. This implied agreement is sometimes called the regulatory compact." Lazar, J. 2016. *Electricity Regulation in the US: A Guide.* 2nd ed. Montpelier, VT: The Regulatory Assistance Project. https://www.raponline.org/wp-content/uploads/2016/07/rap-lazar-electricity-regulation-US-june-2016.pdf.

[19] NASEM, *The Future of Electric Power*, pp. 118–119.

[20] Bonbright, J.C., A.L. Danielsen, and D.R. Kamerschen. 1988. *Principles of Public Utility Rates*. 2nd ed. Dumfries, VA: Public Utility Reports, Inc.

include not only economic efficiency,[21] equity, and fairness[22] (discussed at length in Chapters 4 and 5 of this report), but also ensuring financial sustainability for the utility,[23] providing rates that are simple to understand and relatively stable over time,[24] and supporting electricity system goals (e.g., reliability and resource conservation).

As utility service has evolved from a unidirectional flow of power from the utility to customers in exchange for their payments for utility service, to the current bi-directional situation where some utility customers may not only consume power from the grid but also produce power and inject it into the electric system, there have been many attempts to adjust some traditional ratemaking principles so that they adapt to the new conditions.[25]

Table 7-1 offers an example of an update of such principles and compares traditional ratemaking principles (by Bonbright and colleagues) with those offered by e-Lab experts

[21] According to the *economic-efficiency* principle, rates should be designed so that they signal to customers the full social marginal costs associated with providing them with service and create incentives for customers to make their own decisions about using electricity in ways that are aligned with short- and long-term costs of the system. (However, rates are generally set based on a utility's cost to serve, which does not usually include externalities such as health and environmental costs of the electricity system, and thus would not be economically efficient.)

[22] The *equity and fairness* principle incorporates several elements. On the one hand, similarly situated customers (e.g., residential customers with basic usage patterns) should face *nondiscriminatory* rate structures, such that they have the same opportunities to purchase electricity service under common rates, terms, and conditions, as compared to other sets of customers with different usage patterns (e.g., residential customers with large seasonal electrical demands, or industrial customers with flat usage patterns). But, on the other hand, *due discrimination* in ratemaking may be needed to ensure that all consumers—even those with low incomes—can access basic electricity service.

[23] This principle addresses the need for the local utility to be *financially sound*, able to recover its costs of providing service to consumers, and capable of attracting capital at reasonable cost (including through a fair rate of return on investment).

[24] The *simplicity* principle has long been a feature of rate design, especially for small electricity customers, and is one of the reasons why states have adopted net metering to compensate residential customers that adopt BTM resources. The principle of rate stability is sometimes called *gradualism*, such that changes in rates should be introduced gradually over time rather than through changes that are either sudden or shocking to consumers.

[25] See Lazar, J. 2016. *Electricity Regulation in the US: A Guide.* 2nd ed. Montpelier, VT: The Regulatory Assistance Project. https://www.raponline.org/wp-content/uploads/2016/07/rap-lazar-electricity-regulation-US-june-2016.pdf.

TABLE 7-1 Principles of Public Utility Rate Regulation

e-Lab's Re-interpretation of Bonbright's Principles of Public Utility Rates	
Bonbright Principles	**21st-Century Interpretation**
Rates should be practical: simple, understandable, acceptable to the public, feasible to apply ... and free from controversy in their interpretation.	The customer experience should be practical, simple, and understandable. New technologies and service offerings that were not available previously can enable a simple customer experience even if underlying rate structures become significantly more sophisticated.
Rates should keep the utility viable, effectively yielding the total revenue requirement and resulting in relatively stable cash flow and revenues from year to year.	Rates should keep the utility viable by encouraging economically efficient investment in both centralized and distributed energy resources.
Rates should be relatively stable such that customers experience only minimal unexpected changes that are seriously adverse.	Customer bills should be relatively stable even if the underlying rates included dynamic and sophisticated price signals. New technologies and service offerings can manage the risk of high customer bills by enabling loads to respond dynamically to price signals.
Rates should fairly apportion the utility's cost of service among consumers and should not unduly discriminate against any customer or group of customers.	Rate design should be informed by a more complete understanding of the impacts (both positive and negative) of distributed energy resources on the cost of service. This will allow rates to become more sophisticated while avoiding undue discrimination.
Rates should promote economic efficiency on the use of energy as well as competing products and services while ensuring the level of reliability desired by customers.	Price signals should be differentiated enough to encourage investment in assets that optimize economic efficiency, improve grid resilience and flexibility, and reduce environmental impacts in a technology-neutral manner.

SOURCE: Electricity Innovation Lab, 2014, "Rate Design for the Distribution Edge: Electricity Pricing for a Distributed Resource Future," Mountain Institute, https://rmi.org/insight/rate-design-for-the-distribution-edge-electricity-pricing-for-a-distributed-resource-future.

at the Rocky Mountain Institute.[26] These long-standing rate-design principles, informed by an updated interpretation of the future grid's conditions, are relevant for understanding the balance of considerations needed to shape the evolution of net metering and successor policies.

In the context of net metering, traditional ratemaking principles can create tension across objectives, depending on a number of considerations. For example, as BTM technologies with declining costs provide customers with the opportunity to better manage their electricity bills with on-site generation and those customers take advantage of net metering, the utility may not recover from that customer all of the costs to serve them under some net metering rate designs and service-territory conditions.[27] In that case, the ratemaking goal of "revenue sufficiency" will mean that other customers will have to pick up some of the costs to serve the customer who has BTM generation (see Chapter 4). Such an outcome could raise equity and fairness concerns for non-participating customers, particularly if they tend to be otherwise historically disadvantaged, such as lower income (see Chapter 5).[28] The timing and character of ratemaking reforms to address these outcomes will be influenced in many instances by the principles of simplicity and gradualism, and other considerations (e.g., the extent to which utility rates capture environmental externalities of electricity production).

Principles of Retail Regulation of Utility Payments for Power Supply

As noted above, the adoption of BTM resources creates circumstances in which the customer that adopts the technology ends up producing more power than its on-site electricity consumption and exports power to the grid. In effect, the utility is purchasing such power supply and compensates the customer/supply for those exports. State

[26] See p. 38 in Glick, D., M. Lehrman, and O. Smith. 2014. *Rate Design for the Distribution Edge: Electricity Pricing for a Distributed Resource Future*. Boulder, CO: Electricity Innovation Lab, Rocky Mountain Institute. https://rmi.org/wp-content/uploads/2017/04/2014-25_eLab-RateDesignfortheDistributionEdge-Full-highres-1.pdf. Table 7-1 appeared in Tierney, S.F. 2016. *The Value of "DER" to "D": The Role of Distributed Energy Resources in Supporting Local Electric Distribution System Reliability*. Denver, CO: Analysis Group, Inc. https://www.analysisgroup.com/globalassets/content/news_and_events/news/value_of_der_to-_d.pdf.

[27] Such a situation might occur where the following combination of conditions exists: where the utility's retail rate design recovers a significant portion of the utility's fixed costs through charges that vary with customer usage (e.g., energy [per-kWh charges]), where the variable energy charge is higher than the utility's avoided cost, and where other charges (e.g., customer charge, any capacity or grid access charge, and back-up charges) that are not part of the tariff and possibly are not set at cost.

[28] If the BTM DG does not provide benefits to the system as a whole. Additionally, increasing the price of electricity could work counter to a jurisdiction's electrification goals if it has them.

regulators typically review and approve the terms and conditions under which a utility purchases power from third parties.

Utilities rely on various resources to supply electricity to consumers: from the large and small generating facilities they own; through contracted-for power supply obtained through competitive procurements; through as-available PURPA power supplies; and through resources located on customers' premises that export power.[29] With regulatory approval, supplies of electricity thus tend to be paid for through either: cost-based approaches (e.g., for utility-owned assets), market-based prices (e.g., prices arrived at in competitive procurements, or competitive wholesale markets), or set at the utility's avoided cost (e.g., for PURPA purchases).[30]

In the committee's judgment the utility's compensation to suppliers is thus tied to assets owned by the utility or supplied by an independent non-utility party.[31] In the latter case, the utility's compensation to the power supply tends to be tied to the attributes of the supply and their value to the utility: whether the energy is dispatchable or supplied on an as-available basis (with the former often valued more highly); whether the resource can provide black start or other ancillary services; whether the utility gets the benefit of relying on the resource's capacity when it is of value to the electric system; whether the facility provides other energy services (such as ability to adjust output in response to changes in utility demand, or the ability to provide local reliability value on the transmission or distribution system); whether the resource has environmental attributes such as renewable energy credits (RECs) or zero-emissions credits (ZECs) that are sold to the utility (at a price) or retained by the facility owner. Often, such electricity services (i.e., electrical energy and capacity, RECs, and ZECs) are supplied at market-based prices or paid for at avoided cost.

[29] Rea, J., and O. Ugwu-Oju. 2022. "Why Purchased Power Is Such a Big Deal to Better Understand US Utilities' Climate Impact." https://rmi.org/why-purchased-power-is-such-a-big-deal-to-better-understand-us-utilities-climate-impact.

[30] Cleary, K., and K. Palmer. 2022. *US Electricity Markets 101*. Washington, DC: Resources for the Future. https://www.rff.org/publications/explainers/us-electricity-markets-101.

[31] Allen, N., A. Clair, M. McNaul, R. Shelton, and J. Williams. 2021. *PURPA: Title II Compliance Manual 2.0*. Thompson Coburn LLP. https://pubs.naruc.org/pub/47AD30DC-1866-DAAC-99FB-975A60906D6B; Edison Electric Institute. 2000. "Master Contract." April 25. https://www.eei.org/resources-and-media/master-contract.

THE UTILITY'S OBLIGATION TO SERVE AND ITS RELATIONSHIP TO CUSTOMERS THAT ADOPT TECHNOLOGIES THAT PROVIDE PART OF THEIR OWN ELECTRICITY NEEDS

In the traditional utility business, legal, and regulatory model, a single electric distribution utility enjoys the exclusive right to provide service within the boundaries of a given service territory (sometimes called the franchise area) in exchange for that utility having the requirement to provide electricity service to any and all existing and new customers within that service territory. This combination—of the obligation to serve and exclusive franchise conditions—applies to both investor-owned utilities and publicly and member-owned utilities. All types of utilities operate with the long-standing expectation[32] that rates will be set to recover the cost of providing service to customers. Investor-owned utilities also expect the opportunity to earn returns allowed by the regulator. This obligation to serve at predictable prices or rates is a foundational principle of the standard regulated monopoly utility model.[33]

The advent of BTM DG technologies affords electricity customers the option to have some or all of their electricity supply provided by some entity (e.g., themselves) other than the local distribution utility or a power marketer. This opportunity to self-supply using BTM generation gained some prominence several decades ago as large industrial customers installed cogeneration systems that could provide power and steam to their industrial facilities, and in effect bypass the local utility for some of that customer's electricity needs. That technology helped to spawn the development of independent power facilities that could sell some or all of their output into the grid—which in turn was enabled by the enactment of PURPA, which gave QFs the right to sell their output to the utility at its avoided cost. (That movement also spurred the adoption of retail electricity choice in approximately a third of the states.[34]) More recently, net metering policy has similarly supported the ability of a customer to adopt BTM DG technology to self-supply part or all of their demand and to inject surplus power into the grid.

Under the standard utility model with its obligation to serve, the local distribution utility holds the requirement to supply any, and in many instances all, of a customer's power needs whenever that customer's onsite technology is not providing those services.

[32] For additional information see https://emp.lbl.gov/webinar/recovery-utility-fixed-costs-utility.

[33] NASEM, *The Future of Electric Power*, pp. 117–118.

[34] This includes the following states (plus the District of Columbia) that adopted full or partial retail electricity choice: California, Connecticut, Delaware, District of Columbia, Georgia, Illinois, Maine, Maryland, Massachusetts, Michigan, New Hampshire, New Jersey, New York, Ohio, Oregon, Pennsylvania, Rhode Island, Texas, and Virginia. NREL. n.d. "An Introduction to Retail Electricity Choice in the United States." https://www.nrel.gov/docs/fy18osti/68993.pdf.

Unless a customer with on-site generation (and/or storage) fully and completely disconnects from the grid, the utility continues to face the requirement to plan for, invest in, and operate the system to meet whatever needs those customers have from the grid.[35]

This latter condition means that the utility incurs costs to serve the customer with on-site generation; this arises where the customer relies on the grid to import power when needed or to export power when it exceeds on-site loads, as occurs with net metering. Under the standard utility cost-recovery model, rates are set to cover the cost of the utility's prudent[36] investments and expenses to provide that service. Under standard rate design principles, rates to a particular set of customers are designed to collect the costs incurred to serve those customers.

But, as illustrated in the example described above and discussed in Chapter 4, circumstances may exist[37] where net metering customers' payments to the utility do not cover the costs to provide them with electricity service, and those costs may need to be recovered from other customers. In such situations, there is an asymmetrical obligation to serve and obligation to pay, where the utility has to be ready to serve whenever the customer needs or wants to either withdraw power from the grid or inject surplus electricity into the grid from the BTM resources. This is one of the reasons why, at least in utility service territories with relatively high levels of deployment of rooftop solar, there has been pressure from low-income consumer advocates in some instances, and from utilities, to reform rates for customers with on-site generation so that they pick up more, if not all, of the utility's costs to provide them with service (see discussion in Chapters 3 and 4).[38]

[35] Note that in many jurisdictions where customers with cogeneration facilities have the ability to generate their own power, the utility is allowed to charge standby and back-up rates to cover the utility's cost to be ready to serve those customers' needs when their own generation is not available.

[36] In traditional electric utility regulation, the concept of "prudence" is a key ratemaking concept and refers to a standard of review to determine whether a utility's investments and expenditures may be recovered in rates charged to consumers. A prudency determination examines whether the utility's costs were reasonable at the time they were incurred, in light of the circumstances and what was known or knowable at the time.

[37] This could occur, for example, where the utility's retail rate design recovers a significant portion of the utility's fixed costs through variable energy charges, where those charges are higher than the utility's avoided cost, and where other charges are missing or not set at costs, or both. There are also circumstances in some places where the customer taking service under net metering ends up paying the full cost the utility incurs to provide it with service.

[38] Note that customers with BTM DG who participate in net metering may be providing benefits for all customers (e.g., reducing carbon and pollutant emissions) but these externalities are generally not factored into or able to offset costs utilities incur.

CONSIDERATIONS FOR CONTROL OF NET METERING CUSTOMERS' DER ASSETS

The existence of customer-sited generation, which may be facilitated or supported by net metering, with the potential to export power to the grid poses ownership and control challenges while also enabling new ways to manage and supply the grid.

Using rooftop solar, with and without storage, as an illustrative example, consider the following: The equipment on the roof and the storage system may be owned by an entity different from the customer, including the building owner, the utility, or an independent vendor that has the approval to install the equipment on the premises, with a contractual relationship to the building owner. The BTM equipment interconnects to a distribution system owned by the utility. Under net metering arrangements in most jurisdictions to date, a customer with a BTM DG resource has the option to rely on that resource when it is available and to call on power from the grid when it is not. Depending on the terms of the tariff through which the electricity customer buys electricity service from the utility, the customer (or the owner of the BTM equipment) has ultimate control when it injects power produced or stored into the local grid unless they cede it to the utility.[39]

A project where the customer prefers (and has the technical means) to maintain full control over when to draw power from the grid and when to inject it into the grid may have different operational and cost implications for the utility's operations than a project where the customer has an agreement with the utility for the latter to control when to inject energy into the grid. The former maintains the customer's private option value to use the grid at will; the latter may limit some of the customer's own private value proposition in owning or hosting DG equipment but increases the value proposition to the utility and its other customers (thus meriting greater compensation to the customer or equipment owner, which can offset any loss of private value).

The interactions among (1) BTM technologies and the local distribution system, (2) the customer's ability to determine whether they maintain the ability to control exports to the system at different times and levels of output, and (3) the need for the utility to maintain reliable and resilient service to all of its customer base, together mean that the utility may need to invest in communications and control technologies on the distribution

[39] In most jurisdictions, the utility maintains the ability to disconnect the customer's equipment to prevent it from injecting power into the grid for emergency safety and potentially other reasons. Note that there is some interchangeability between direct utility control over DG assets located on the customer's premises and price signals that either incentivize or commit the customer to operate its system in the same manner. The most clear-cut example is direct load control programs versus DR call options (in which the utility has a service agreement with a customer that allows the utility to call on the customer to provide the demand reduction when a particular price or other condition occurs).

system so that the utility has visibility into conditions on its system. Regulators, in turn, will need to view these investments as critical elements of the transitions occurring at the grid edge and in many cases enabled by net metering policy, and approve recovery of their costs (see further discussion in Chapter 6).

INTERACTIONS OF NET METERING WITH LOCAL REGULATORY POLICIES AND CONDITIONS

As described in Chapter 3, the pace of deployment of DER across the country varies considerably, in part due to differences in local conditions affected by state regulatory policies, the structure of the electric industry in the state, the current generation mix, the maturity of the local DG market, and other drivers such as the quality of solar resources, the price of electricity,[40] and vulnerability to extreme weather events. Of these, legislative and regulatory directives to implement net metering policies—or to block them—play significant roles.

For example, the corporate character of electric utilities can sometimes play a role in either impeding or enabling the adoption of both net metering as well as BTM DG and other DER, such as rooftop solar. Most end-use and retail electricity customers in the United States are served by an investor-owned utility (IOU) whose rates are subject to regulation by state public utility commissions even though most local utilities are actually publicly owned—because they are either part of a local municipal government or a member-owned cooperative electric utility, or even a special-purpose utility district (e.g., an irrigation district) that evolved historically to supply power to local constituents.[41] Most of these publicly owned utilities provide only distribution service, with their power supplied by publicly owned generation and transmission utilities (e.g., the Tennessee

[40] For example, among states that offer net metering at or close to retail rate, rooftop solar adoption per capita is higher in states with higher electric rates. See Davis, L. 2020. "Putting Solar in the All the Wrong Places." *Energy Institute Blog*. Berkley, CA: Energy Institute at HAAS. https://energyathaas.wordpress.com/2020/02/03/putting-solar-in-all-the-wrong-places.

[41] "In 2018, even though there are fewer than 200 investor-owned utilities (compared to approximately 2,900 publicly owned utilities (including cooperatives, municipal utilities, and special-purpose utility districts), approximately three-quarters of all retail power went through the local distribution systems of these investor-owned utilities (EIA 2019a,b). Because cooperatives serve rural areas, the geographic footprint of their systems is often large ... and more of their sales are to residential customers (over 50%) as compared to other distribution utilities (whose residential sales account for about a third of all retail sales)." (Footnotes in the original have been omitted.) EIA (U.S. Energy Information Association). 2019a. "861 Electric Data." https://www.eia.gov/electricity/data/eia861; EIA. 2019b. "Electric Sales, Revenue, and Average Price." October. https://www.eia.gov/electricity/sales_revenue_price; NASEM, *The Future of Electric Power*, pp. 96–98.

Valley Authority and Tri-State Generation and Transmission Company) or investor-owned utilities.

As explained in the National Academies' 2021 study *The Future of Electric Power in the United States*, the varied institutional, financial, and governance structures of these many types of utilities create different systems of incentives and constraints on their willingness and ability to innovate as they carry out their public service obligations.[42]

For example, an IOU with a traditional rate design that collects the utility's fixed costs (including return of and on investment) largely through variable energy charges and whose rates are set through traditional general rate cases without revenue decoupling will likely experience erosion of electricity sales and revenues with customers' adoption of BTM generation. This may position that utility's financial interests in tension with customers' interests in the adoption of BTM, and this tension may be exacerbated where a net metering policy compensates those customers' exports at a rate higher than the utility's avoided costs. Under these circumstances, and all else equal, this utility will lose revenues and profits until rates are reset in another rate case (see further discussion in Box 7-1).

Under a similar set of assumptions, a publicly owned utility will not face the exact same financial outcomes (because it has neither shareholders nor profits), but even that utility could face financial tensions with lost sales and lost revenues resulting from net-metered BTM generation: All else equal,[43] that utility must eventually collect sufficient revenues to cover its costs, and must adjust rates to all of its customers over time to assure revenue adequacy.[44]

A second dimension which might affect utilities' appetite for customers' adoption of BTM DG through the design of net metering policies and their potential impact on cost recovery is the structure of the electric industry combined with rate design in a particular place. Using the prior example, an IOU that is vertically integrated—owning generation, transmission, and distribution facilities—*and* has rates that are not subject to revenue decoupling and recovers fixed costs substantially through variable energy charges, will experience revenue erosion for its entire bundled rate. By contrast, an IOU that provides only "wires" service to its customers will not be exposed to erosion of revenues associated

[42] NASEM, *The Future of Electric Power*, p. 93.

[43] Electrification of some building end uses and vehicles over time could be expected to increase sales of electricity and may mitigate the timing or need to adjust rates to address these lost revenues from BTM generation.

[44] Also, many electric distribution coops have long-term contracts for generation and transmission service from another coop, in which there may be terms and conditions that restrict the former's savings on variable costs.

BOX 7-1
NET METERING, UTILITY BUSINESS MODEL, AND RATE DESIGN

Many observers have noted reasons why utilities might be concerned about BTM resources' potential adverse impacts on the utility's revenues and future investment opportunities. Under the traditional investor-owned utility business model, utilities are driven to increase sales and to increase capital investments on which they may collect a return on investment.

Retail rates for many customers have been designed in a way that ends up recovering large parts of what are considered fixed costs (e.g., distribution grid infrastructure) through volumetric charges per kWh of sales. This general trend has made the volume of sales central to utility revenue collection and profitability in the traditional model. Increasing adoption of net-metered BTM DG can reduce utility sales and thereby revenues. Shortfalls in revenues must then be spread over a smaller customer base, raising average retail rates, thus creating both distributional impacts, but also additional incentives for more customers to adopt BTM DG or otherwise reduce their electricity consumption from the utility. At the extreme, this feedback loop has been referred to, somewhat dramatically, as the "utility death spiral." This dynamic is the basis for the general observation that growth of net-metered BTM DG is misaligned with the traditional utility business model, which in turn has led to the consideration of a number of potential adjustments to net metering policies as well as changes to the utility business models to address this underlying misalignment.

One way to address the misalignment is through the rate structure itself wherein a higher fixed charge is levied on BTM DG customers or customers generally: with more revenue collected through higher fixed monthly charges, the utility revenue would not depend so heavily on volume of sales, and thus reduced sales due to higher BTM DG would not reduce revenues (as much) to utilities. However, high fixed charges can limit customers' ability to manage their bills (including through energy efficiency since a high portion of it may be fixed), raising concerns for low- and fixed-income customers (Satchwell et al. 2015). The imposition of high fixed charges could also theoretically encourage some customers to disconnect from the grid altogether and actually exacerbate the utility's problem of loss of sales.

Another way to address the misalignment might be to consider use of other net metering variants (e.g., net billing or compensation for exports at less than the full retail rate, or buy-all/sell-all rates that otherwise vary across inflows and outflows of power).

A third approach to better aligning utility incentives with BTM DG includes changes to regulation related to utility cost recovery to support or encourage changes to the traditional utility business model focused on increasing sales. In particular, insights from the range of regulatory tools that have been used to correct for similar misalignments in the case of energy efficiency (EE) are also potentially applicable in the case of BTM DG. Two of the most relevant mechanisms include *revenue decoupling* and *lost revenue adjustment mechanism* (LRAM). Revenue decoupling breaks the link between utility revenues and sales, thus enabling utilities to collect allowed revenues regardless of changes in sales. LRAM is more narrowly focused on directly adjusting utility revenues that can be lost due to a utility program (e.g., an EE measure). While decoupling and LRAM mechanisms may address the utility's declining revenues problem, they can also increase a shift in costs from customers adopting BTM DG and participating in net metering to non-participants.

continued

BOX 7-1 Continued

A fourth approach could consider utility ownership of BTM DG, thus bringing BTM DG investments by utilities within the fold of traditional rate-based utility business model. Under this approach, utilities lease rooftops from customers, but own the PV systems—and thus earn a rate of return on those investments. In theory, this approach aligns the utility's financial interests with those customers seeking to have power supplied by solar resources, but it is not clear whether such an approach would pass the current regulatory requirement to demonstrate that the investment is cost-effective for customers generally and the distributional impact on DG versus non-DG customers is equitable. It is also likely mediated by a number of contextual and infrastructure factors. See, for example, Barbose and Satchwell (2020): "Whether or not regulators would find utility-owned rooftop solar to be in the public interest may therefore be conditional on its ability to capture those additional sources of value and to advance other policy goals, such as emission reductions, customer choice and improving solar access among low- and moderate-income communities." In this context, regulators may want to consider whether there are ratemaking approaches that align the utility's financial interests in adoption of technologies with society's or customers' interests in non-polluting sources of generation (whether centralized or on their customer's rooftops).

References

Altunay, M., A. Bergek, and A. Palm. 2021. "Solar Business Model Adoption by Energy Incumbents: The Importance of Strategic Fit." *Environmental Innovation and Societal Transitions* 40:501–520.

Barbose, G., and A.J. Satchwell. 2020. "Benefits and Costs of a Utility-Ownership Business Model for Residential Rooftop Solar Photovoltaics." *Nature Energy* 5(10):750–758.

Blackburn, G., C. Magee, and V. Rai. 2014. "Solar Valuation and the Modern Utility's Expansion into Distributed Generation." *The Electricity Journal* 27(1):18–32.

Davies, L.L., and S. Carley. 2017. "Emerging Shadows in National Solar Policy? Nevada's Net Metering Transition in Context." *The Electricity Journal* 30(1):33–42.

Funkhouser, E., G. Blackburn, C. Magee, and V. Rai. 2015. "Business Model Innovations for Deploying Distributed Generation: The Emerging Landscape of Community Solar in the US." *Energy Research & Social Science* 10:90–101.

Kubli, M. 2018. "Squaring the Sunny Circle? On Balancing Distributive Justice of Power Grid Costs and Incentives for Solar Prosumers." *Energy Policy* 114:173–188.

Laws, N.D., B.P. Epps, S.O. Peterson, M.S. Laser, and G.K. Wanjiru. 2017. "On the Utility Death Spiral and the Impact of Utility Rate Structures on the Adoption of Residential Solar Photovoltaics and Energy Storage." *Applied Energy* 185:627–641.

O'Shaughnessy, E., and K. Ardani. 2020. "Distributed Rate Design: A Review of Early Approaches and Practical Considerations for Value of Solar Tariffs." *The Electricity Journal* 33(3):106713.

Riveros, J.Z., M. Kubli, and S. Ulli-Beer. 2019. "Prosumer Communities as Strategic Allies for Electric Utilities: Exploring Future Decentralization Trends in Switzerland." *Energy Research & Social Science* 57:101219.

Satchwell, A., A. Mills, and G. Barbose. 2015b. "Regulatory and Ratemaking Approaches to Mitigate Financial Impacts of Net-Metered PV on Utilities and Ratepayers." *Energy Policy* 85:115–125.

with some or all of the generation portion of the customers' rates (because that portion of their electricity service is supplied by a retail marketing company or default service where the utility passes through the costs it incurs).[45] The loss of profits will nonetheless may be viewed as a disincentive to customer-adoption of BTM generation (and other DER).

Wires-only IOUs operate in about a third of the United States. (Most rural electric cooperatives also only own and operate distribution facilities and buy power on an all-requirements basis from third parties, which often includes a generation-and-transmission cooperative.)[46] Thus, a utility's attitude toward net metering is a function of many factors, including the potential impact of net metering on the utility's earnings and the impact of net metering on non-participating customers (if their rates need to be adjusted to recover lost revenues).

As noted, net metering does not directly impact earnings, or profits (versus cost recovery), for all utility types. Publicly owned and municipal utilities are not profit-driven, although they do need to be able to finance investment and recover their costs. Not all IOUs earn profits through energy sales due to the wide adoption of revenue decoupling which is a policy mechanism through which IOU earnings are made independent of (or decoupled from) the amount of electricity they sell.[47] IOUs that do not have decoupled rates[48] are more likely to view net metering-enabled exports from BTM generation negatively because they directly impact electricity sales from utility-owned generation assets.

Profits for revenue-decoupled utilities, such as IOUs in California, Massachusetts, and New York, are not directly affected by the amount of energy sales through net metering. The profits for these companies may be indirectly affected to the extent net

[45] Also, in states with retail choice, net metering rates often credit the customer's exports to the grid at the generation portion of the overall rate, which further dampens the revenue erosion for the distribution utility.

[46] EIA 861 data, Sales_Utl_Customer_2021.xlsx. That spreadsheet allows for toggling, and when you toggle "service type" to show delivery only (i.e., wires) and toggle "ownership" to show investor owned, you get the following states identified (in column g): California, Connecticut, Delaware, District of Columbia, Illinois, Maine, Maryland, Massachusetts, Michigan, Montana, Nevada, New Hampshire, New Jersey, New York, Ohio, Oregon, Pennsylvania, Rhode Island, Virginia, and Washington. EIA (U.S. Energy Information Association). 2019a. "861 Electric Data." https://www.eia.gov/electricity/data/eia861.

[47] As of 2019, approximately half of the states had adopted revenue decoupling for at least some if not all of their electric utilities. National Governors Association. n.d. "State Energy Efficiency Policy in a New Era: A Toolkit for Governors." https://www.nga.org/wp-content/uploads/2021/10/Energy-Efficiency-Toolkit.pdf.

[48] As of 2019, states with IOUs and without decoupling for electric utilities included Florida, Kentucky, Louisiana, Mississippi, Missouri, Montana, Nebraska, New Mexico, North Dakota, Oklahoma, South Carolina, South Dakota, and West Virginia. See p. 4 in Cleveland, M., L. Dunning, and J. Heibel. 2019. "State Policies for Utility Investment in Energy Efficiency." National Conference of State Legislatures. https://www.ncsl.org/Portals/1/Documents/energy/Utility_Incentives_4_2019_33375.pdf?ver=2019-04-04-154310-703.

metering-enabled distributed generation changes the amount of grid infrastructure investments a utility needs to make, in one direction or another. For example, but for the deployment and operation of this DG resource (which could be enabled by net metering and/or other regulatory policies), the distribution utility would need to make investment on which it would be able to earn a return. In other circumstances, net metering-enabled BTM generation could trigger the need for distribution-system or infrastructure upgrades with the IOU having the opportunity to invest in and earn a regulated rate of return on that new infrastructure.[49]

Estimating whether and to what extent net metering-enabled BTM generation impacts the need for infrastructure upgrades is location—and time-dependent, such that one cannot conclude a priori that it will create tensions with an IOU's own financial incentives. Additionally, the design of any net metering variant, in combination with other regulatory tools (e.g., identification of available or constrained hosting capacity for DGs on distribution systems or policies that enable a utility to share in any avoided distribution-system costs) may affect such incentives as well, in positive or negative ways.

DG can provide benefits to the system, and thereby reduce costs, by enhancing system resilience[50] and reducing externalities associated with electricity production (as explained in Chapters 4 and 5). This consideration should be taken into account in designing net metering variant policies to compensate the supplier of BTM resources for any such resilience value supplied or externalities avoided, thus providing additional benefits to the grid and other customers or avoiding the utility's alternative investments and associated costs to achieve that outcome.[51]

However, DG can introduce costs to the system for the interconnection and integration of many more devices to the grid (as discussed in Chapter 6). Such costs may include the additional controls needed to be added to the operations of the local distribution system and increased risks of cyber-attacks. These costs may increase as the penetration rates of

---

[49] See Technology Chapter 6. See also Wolak, F.A. 2018. *The Evidence from California on the Economic Impact of Inefficient Distribution Network Pricing*. Cambridge, MA: National Bureau of Economic Research. https://www.nber.org/papers/w25087.

[50] See, for example, Converge Strategies, LLC. 2019. "The Value of Resilience for Distributed Energy Resources: An Overview of Current Analytical Practices." Prepared for the National Association of Regulatory Utility Commissioners. https://pubs.naruc.org/pub/531AD059-9CC0-BAF6-127B-99BCB5F02198; NREL (National Renewable Energy Laboratory). 2018. "Valuing the Resilience Provided by Solar and Battery Energy Storage Systems." https://www.energy.gov/sites/prod/files/2018/03/f49/Valuing-Resilience.pdf; Van Nostrand, J.M. 2019. "Quantifying the Resilience Value of Distributed Energy Resources Quantifying the Resilience Value of Distributed Energy Resources." *Journal of Land Use* 35(1):15–34.

[51] The discussion here focuses on the role of BTM DG to the grid, rather than the value BTM DG provides to the customer that puts it in place. From a customer's point of view, the *private* resilience value of BTM DG (including storage) may be an important consideration in the decision to add it.

DG systems increase. Utility regulators should take into account the full range of positive and negative implications as they consider policies to support BTM DG, to compensate DG suppliers for their exports to the grid, and to allow the recovery of utility investment to integrate increasing amounts of DG into the electricity system. Otherwise, customers may forego the potential benefits of BTM DG to the system and remaining customers may need to bear costs to make sure that the operations of the grid remain reliable.

Utility regulators have a critical role to play in ensuring that net metering (or its variants) provides fair compensation that reflects the value of DG to meet energy needs and relevant policy goals (e.g., resilience and emissions reductions) while also covering the utility's costs to provide service. Because financial considerations may create disincentives for utilities in collecting data and conducting analyses, regulators should make sure that the utility conducts the analyses necessary to determine reasonable estimates of both avoided costs and delivered benefits or make the utility's data available to the public so that others can provide such reasonable estimates.

Additionally, policymakers, utility regulators, and boards of publicly owned utilities that seek to promote BTM generation should also recognize the importance of the distribution utility's investment in grid technology and other enhancements to provide visibility of DG output and other communications and control technologies (as explained further in Chapter 6).

Accounting for Clean Energy and Environmental Mandates in Net Metering

Many states have some form of clean energy or environmental mandates with which utilities need to comply (see Chapter 3 for further discussion).[52] When customers install clean DG on their premises, their power providers have to contract with or build fewer resources that qualify to meet their regulatory clean energy requirements. Therefore, customer-adopted clean DG may provide a benefit to all utility customers by deferring utility procurement of supply-side clean resources in jurisdictions with clean energy

[52] This can arise, for example, where BTM generation consumed onsite reduces the load-serving entity's supply (because the grid sees the customer's demand as having been reduced by the on-site generation) and therefore reduces the amount of renewable generation the supplier needs to arrange in order to meet RPS standards that are based on a percentage of load. In many but not all states, BTM generation—whether consumed onsite or exported to the grid—can qualify for producing renewable energy credits. Those RECs may be owned by the customer or the owner of the BTM system, depending on the terms of contractual arrangements where the customer does not own the system. If the BTM system produces RECs for the full output, then there could be double counting of the renewable generation. States may wish to ensure accounting methods take this possibility into account that appropriate accounting for the net value of such resources.

requirements (or enabling the utility to procure fewer overall clean resources). In jurisdictions where such a benefit exists, regulators (and boards of publicly owned utilities) may reflect this benefit in the compensation arrangements for DG resources.

Regulators thus seeking to assure that compensation for exports of on-site generation are valued properly may do so in ways that are consistent with other ratemaking policies and public policy considerations. For example, in principle, regulators in states with carbon-reduction commitments need to establish the carbon-reduction value of on-site DG consistent with the incremental costs to comply with their state's clean energy policy and/or consistent with subsidies or incentives provided to other sources of zero-carbon electricity supply. Symmetrically, it would be appropriate for policymakers to also charge polluting sources for the carbon and other pollutants they emit. Box 7-2 gives an example of one state's net metering journey.

PRINCIPLES FOR THE DESIGN OF POLICIES THAT AIM TO LAUNCH AND INVIGORATE MARKETS FOR NEW TECHNOLOGIES

The history of net metering indicates the important role that it has played in helping to support customers' adoption of BTM technologies and to build a market for rooftop solar, in particular, where it had previously not existed. Since the first formal net metering program was put in place nearly 40 years ago,[53] conditions in the market for renewable BTM technologies have evolved substantially, as described in detail in the Chapter 3. The installed costs of solar panels (both rooftop and supply side) have dropped 60 percent over the past decade,[54] due in part to demand driven by states' renewable portfolio standards (RPS) and net metering compensation arrangements, manufacturing efficiencies, and other factors. The benefits of these economic improvements have accrued not just in the states that were early adopters of net metering, but also in other states as well. Additionally, net metering has been especially successful in stimulating customer adoption of BTM technologies in the states where retail prices were high and net metering

[53] Minnesota was the first state to enact a net metering program in 1983. Solar Electric Power Association. 2013. "Ratemaking, Solar Value and Solar Net Energy Metering—A Primer," pp. 5–6. Washington, DC: Solar Electric Power Association. https://www.energy.gov/sites/prod/files/2015/03/f20/sepa-nem-report-0713-print.pdf.

[54] Solar Energy Industries Association. 2022. "Solar Industry Research Data." https://www.seia.org/solar-industry-research-data. Wood Mackenzie and Solar Energy Industries Association (SEIA) indicate that during 2022, there was an uptick in the pricing of installed solar systems to all customer segments due to a variety of factors (e.g., international trade considerations and price increases in polysilicon and other raw materials). Wood Mackenzie and SEIA. 2022b. "U.S. Solar Market Insight Report: Executive Summary—Q3 2022," p. 16. https://www.woodmac.com/industry/power-and-renewables/us-solar-market-insight.

BOX 7-2
CALIFORNIA AND NET METERING

Over the past 25 years, California has adopted multiple statutory and regulatory directives that have directly or indirectly affected its net metering program, and the state's policy support for BTM generation.

California first adopted its original net metering program as part of a new law, SB 656, in 1996. Citing the purpose of providing net metering to "encourage private investment in renewable energy resources, stimulate in-state economic growth, enhance the continued diversification of California's energy resource mix, and reduce utility interconnection and administrative costs," the program (eventually known as "net metering 1.0") required every electric utility in the state to establish a tariff or contract to enable net metering for relevant residential customers and to make it available on a first-come, first-serve basis up to the point when customer-sited generators equaled 0.1 percent of the utility's peak electricity demand forecast for 1996.[a]

In 2013, California's Assembly Bill (AB) 327 increased the cap on net metering-customer additions and directed state regulators to develop an updated net metering tariff (informally known as "net metering 2.0") which would be available to customers wanting to install rooftop solar after the net metering 1.0 cap was reached. AB 327 also imposed minimum bills for net metering customers and mandated rate reforms to mitigate cost-shifts between net metering customers and other customers (including prohibiting cost shifts across customer classes).

Since 1996, penetration rates for residential customers with rooftop PV and on net-metered rates increased to 11.4 percent as of May 2022. California leads the country in total BTM solar capacity installations. (EIA, Form EIA-861M, detailed data.[b])

The California Public Utility Commission (CPUC) opened a proceeding in 2020 to consider potential changes to net metering 2.0. This proceeding became a highly contested docket with review of such issues as: the existence and size of any cost shift from net metering participants to non-participants; the methodology for calculating the value of exports from net metering customers' on-site systems to the grid; the question of whether net metering 1.0 and 2.0 customers should be grandfathered; the appropriate length of a transition period to move customers off their current net metering service; provisions relating to the netting period, rate design, addition of any grid or other charges impacts on low-income customers; calculations of bill savings and payback periods for customers adopting behind-the-meter resources under net metering tariff options; impacts on the rate of adoption of solar PV; and whether to provide particular incentives to encourage the adoption of storage in conjunction with solar systems or incentives targeted for low-income customers in particular.

On December 15, 2022, the CPUC adopted a successor tariff to net metering 2.0 for new customers adopting eligible BTM technologies.[c] The CPUC described that the adopted tariff:[d]

- "Applies new residential rates to encourage electricity use when it is most beneficial for grid reliability. These rates have significant differences between peak and off-peak prices to incent battery storage and load shifting from evening hours to overnight or midday hours."

continued

BOX 7-2 Continued

- "Credits solar and solar plus battery storage customers for the electricity they export to the grid based on its value, as determined by the avoided cost to their utility of buying clean electricity elsewhere."
- "Provides extra electricity bill credits to residential customers who adopt solar or solar paired with battery storage in the next five years, which are paid on top of the avoided cost bill credits. Customers are guaranteed these extra bill credits for 9 years."
- "Expands access to solar and storage for low-income customers, residents living in disadvantaged communities, and residents living in California Indian Country by providing a larger amount of extra bill credits."
- "Increases the allowable size of rooftop solar systems to cover 150 percent of a customer's electricity usage to accommodate future electrification of appliances and vehicles."

---

[a] Eligible customers had to own and operate a solar system not more than 10 kW in size, located on the customer's premises and intended primarily to offset part or all of the customer's own electrical requirements. "'Net energy metering' means using a single, nondemand, non-time-differentiated meter to measure the difference between the electricity supplied by a utility and the electricity generated by an eligible customer-generator and fed back to the utility over an entire billing period." California Legislative Information. 1995. "An Act to Add Section 2827 to the Public Utilities Code, Relating to Public Utilities." State of California Senate Bill No. 656. http://www.leginfo.ca.gov/pub/95-96/bill/sen/sb_0651-0700/sb_656_bill_950804_chaptered.html.

[b] This calculation is based on (1) May 2022 data on net-metered data for California by customer class (which total 1,462,369 for residential customers in CA) and (2) May 2022 data on sales to ultimate customers by customer class in California (which totaled 12,849,402 for residential customers in CA). EIA. 2022d. "Electricity: Form EIA-861M (formerly EIA-826) Detailed Data." https://www.eia.gov/electricity/data/eia861m.

[c] CPUC. 2022d. "Order Instituting Rulemaking to Revisit Net Energy Metering Tariffs Pursuant to Decision 16-01-044, and to Address Other Issues Related to Net Energy Metering." Decision proposed to CPUC. Rulemaking 20-08-020. https://docs.cpuc.ca.gov/PublishedDocs/Published/G000/M499/K921/499921246.PDF.

[d] CPUC. 2022a. "CPUC Modernizes Solar Tariff to Support Reliability and Decarbonization," December 15, 2022. https://www.cpuc.ca.gov/news-and-updates/all-news/cpuc-modernizes-solar-tariff-to-support-reliability-and-decarbonization.

---

rate design provided an attractive financial incentive to install residential photovoltaic (PV) systems.

Thus, some of the original rationales for the adoption of traditional net metering—such as animating customer demand to help increase private investment in BTM technologies, help drive down the costs of rooftop solar, and help build the supply-side of the marketplace—have been demonstrably successful in some parts of the country.

This success after decades of experience provides an opportunity to assess the conditions under which policymakers can design second- or third-generation successor tariffs in places where net metering has been in place for years and can fashion variant or alternative approaches where markets are less mature (see Chapter 3 which provides insights into the ways in which net metering and its variants have adapted to changing conditions over time). As with traditional regulation of utilities, one of the consequences of putting a policy incentive in place is that the constituencies that have benefitted from the current incentive become accustomed to its presence, with utilities and firms sometimes building business models designed around the incentive and with customers expecting to continue to enjoy the incentive even when changed conditions may no longer warrant the continued availability of the incentive in its original form.[55]

The literature[56] on the use of economic incentives in public policies to stimulate market development for the adoption of new technologies suggests conditions under which a policy should be permanent versus "transient." In other words, policy design (e.g., for a policy like net metering) might take into account conditions in the market and the appropriateness of temporary versus permanent incentives to encourage demand for and adoption of a technology over time.[57] Technology learning curves are relevant for aligning policy design with the accomplishment of policy objectives over time, and for identifying appropriate policy instruments—such as net metering versus net metering successor tariffs versus alternative funding mechanisms (e.g., programs supported by public budgets rather than electricity rates)—to accomplish the policy goals.[58]

The National Academies' Committee on the Future of Electric Power in the United States made this point in discussing the importance of designing policy so that it evolves over time as conditions change:

> Policy makers and stakeholders should be ready to modify policies as conditions evolve, including through the use of sun-setting mechanisms in policy to phase them out over time.

[55] See, for example, NASEM, p. 139, "Reducing Reliance on Less-Efficient Policies May Face Opposition," in *The Future of Electric Power*; Stavins, R.N., T. Schatzki, and R. Scott. 2019. "Transitioning to Long-Run Effective and Efficient Climate Policies." *M-RCBG Faculty Working Paper Series* 2019-01. https://www.hks.harvard.edu/sites/default/files/centers/mrcbg/files/FWP2019-01.pdf.

[56] This literature sometimes refers to the role of technology adoption, economic and market learning curves, and "learning by doing." See, for example, Speelman, L., and Y. Numata. 2022. "Harnessing the Power of S-Curves: How S-Curves Work and What We Can Do to Accelerate Them." https://rmi.org/insight/harnessing-the-power-of-s-curves.

[57] See, for example, Gillingham, K., and J.L. Sweeney. 2010. "Chapter 5: Market Failures and the Structure of Externalities," in *Harnessing Renewable Energy in Electric Power Systems: Theory, Practice, Policy*. New York: Earthscan; also in Researchgate. "Market Failures and the Structure of Externalities," p. 28 of 32, and p. 85, Table 2. https://www.researchgate.net/publication/228345632_Market_Failure_and_the_Structure_of_Externalities.

[58] An example of state regulators evolving their ratemaking instruments over time is the case of energy efficiency and revenue decoupling.

> Often, such policies are adopted based on the expectation that there are barriers to entry of new technologies and that early entrants with a new technology and facing limited markets for their products will have relatively high costs. The policies are often designed to overcome these costs and then to allow growth in the market to push costs down....
>
> Establishing time limits or sunsets for certain market-conditioning policies is important as a way to avoid unintended consequences.[59]

That committee made a finding that is relevant for the future of net metering as a tool for advancing the development of a market for BTM technologies:

> Finding 3.13: More care is needed in the design of policies so that they can sunset or phase down over time when their original purposes have been met and economic, technological and social conditions change.[60]

Also, state policymakers (notably, legislators and governors) should consider the value of using other policy approaches (e.g., investment incentives and other programs supported through taxpayer-funded public budgets,[61] or changes in building codes[62]) rather than electricity rates as a means to advance policy objectives relating to economic and market development, such as incentives to leverage private investment and climate-friendly energy supplies with no greenhouse gas emissions. Ultimately, such goals are societal goals rather than electricity system goals, which merit support by a broad-based funding source—and one with fewer inequitable impacts on low-income electricity customers.

FINDINGS

Finding 7-1: The design of rates for the purchase of electricity by customers with BTM DG and other DER and for compensation for their exports to the grid is subject to pricing decisions of local decision makers (i.e., state regulators for investor-owned utilities and boards of other distribution utilities). To date, relatively few, narrow circumstances exist where FERC regulates compensation or rates for net exports into the grid from behind-the-meter resources.

---

[59] NASEM, *The Future of Electric Power*, p. 129.

[60] Ibid, p. 130.

[61] The 2022 federal Inflation Reduction Act includes tax credits for homeowners that install rooftop solar, or solar + storage. See EERE (Office of Energy Efficiency and Renewable Energy). 2022a. "Homeowner's Guide to the Federal Tax Credit for Solar Photovoltaics." https://www.energy.gov/eere/solar/homeowners-guide-federal-tax-credit-solar-photovoltaics.

[62] California Energy Commission. 2022. "Solar PV Systems and Solar Ready." https://www.energy.ca.gov/programs-and-topics/programs/building-energy-efficiency-standards/online-resource-center/solar.

Finding 7-2: Decision makers have to make ratemaking decisions—about net metering and the design of other rates—that take into account not only economic efficiency and fairness considerations, but also other important principles such as revenue sufficiency for the utility, simplicity of rates, and gradualism in rate changes over time. In some jurisdictions, other important ratemaking considerations include providing rate discounts for low-income customers, clean energy, economic development, and other factors.

Finding 7-3: Because the local distribution utility has an obligation to serve, the utility has to plan for and be ready so that the electricity system can meet the needs of a customer even when that customer has BTM DG and/or storage. For example, the utility has to be able to serve the customer when they use the grid and import power or when BTM power exports to the grid. Thus, the utility incurs costs to serve such needs.

Finding 7-4: Circumstances may exist where net metering customers' payments to the utility do not cover the costs to provide them with electricity service, and those costs need to be recovered from other customers, leading to pressure to reform rates for customers with on-site generation so that they cover more, if not all, of the utility's costs to provide them with service. In other circumstances, the utility's compensation to the customer for its exports to the grid may be lower than the value of the services those exports provide to the grid.

Finding 7-5: Estimating whether and to what extent net metering-enabled BTM generation impacts the need for infrastructure upgrades is location- and time-dependent, such that one cannot conclude a priori that it will create tensions with an IOU's own financial incentives.

Finding 7-6: Utility regulators play a key role in enabling distribution system infrastructure investments where needed to integrate and manage DG technologies and installations by allowing recovery of these costs to capture the benefits of BTM DG for other customers and the system as a whole. Regulatory standards and policies that support the adoption of DG and DER technologies and systems need to be accompanied by regulatory decisions to approve investments to integrate those technologies and enable the grid to operate reliably with them in place. Such investments may take place on the utility's, customers', or non-utility entities' assets.

Finding 7-7: Some of the original rationales for the adoption of net metering—such as animating customer demand to help increase private investment in BTM technologies, to help drive down the costs of rooftop solar, and to help build the supply-side of the marketplace—have been demonstrably successful in some parts of the country where net metering has been adopted.

Finding 7-8: The success of net metering in many states provides an opportunity to assess the conditions under which policymakers can design second- or third-generation successor tariffs (or net metering variants) in places where net metering has been in place for years and can fashion alternative approaches where markets are less mature. Good policy design should build in opportunities to change incentives when objectives are met or change.

Finding 7-9: Rather than using a ratemaking tool like net metering or other programs whose costs are recovered in electricity rates in order to advance public policy goals, policymakers have the option to adopt other policies (e.g., tax-funded incentives and programs) to accomplish important societal goals. Over-reliance on net metering (as compared to other policy tools like taxpayer-funded incentives) to accomplish these objectives may lead to unintended consequences (e.g., cost shifts or lack of access to BTM technologies by low-income consumers). Because electricity rates are generally more regressive than income taxes, using taxpayer funded programs or other policy tools (e.g., regulation or performance standards) may provide more equitable outcomes than using electricity rates to accomplish and pay for these programs.

RECOMMENDATIONS

Recommendation 7-1: Decision makers about electric utility rates—including state legislators and utility regulators and governing boards of publicly owned electric utilities—should take into account that DG technology costs and market maturity are at a stage, both technically and economically, where traditional net metering policies to support the deployment of DGs need to be assessed and revisited. This recommendation applies both to instances where a utility operates in a state that previously adopted net metering and regulators are considering variants to it, as well as in parts of the country that have not yet adopted net metering and seek to advance BTM technologies, and have the option to leapfrog beyond net metering and adopt other ratemaking variants.

Recommendation 7-2: Decision makers should rely on important traditional ratemaking principles as updated to reflect the application of new technologies and service offerings. These updated principles include cost-causation, rate simplicity,

fairness, revenue adequacy, a simple customer experience even if underlying rate structures become significantly more sophisticated, and compensating resources based on their value.

Recommendation 7-3: Decision makers should design compensation approaches for the export of power from BTM generation according to principles that are consistent with how the utility values other sources of power that offer comparable energy, capacity, and other grid services to the system (which may vary by time and location). External impacts (such as pollution) from some sources of power are unpriced impacts, and in many jurisdictions there are constraints on the ability of regulators to reflect externalities in utility planning and/or ratemaking. Sound economic principles would support the consideration of such externalities in utility regulation; policymakers should consider how to address such impacts in utility and other energy policies.

Recommendation 7-4: Given the economic and equity challenges associated with using net metering—and with financial incentives and programs recovered in electricity rates more generally—to promote investment in and deployment of DG technologies, policymakers should also consider, and where appropriate use, other policy instruments such as tax incentives, building codes, attractively priced loans (or even grants) to low-income households, and other complementary policies.

8

The Future of Net Metering in an Evolving Electricity System

A confluence of changes in technology, policy, economics, and customer expectations have driven fundamental shifts in the electricity system over the last two decades. Amid the global imperative to reduce greenhouse gas (GHG) emissions, innovations in and rapidly declining costs of distributed generation (DG) and distributed energy resource (DER) options have made the demand-side of the electricity system much more dynamic. Customers can choose among more and different energy products and services and can actively manage their electricity use and even generate their own electricity. These changes are precipitating the need for innovation in utility business models and regulatory frameworks to keep pace.

In this context, net metering has played an important role in increasing the deployment of behind-the-meter (BTM) DG. Distributed solar has proliferated over the last two decades to a degree that may not have been easily imagined 20 years ago. Net metering, a billing mechanism to credit and compensate customers for production from clean BTM DG that offsets consumption and is exported to the grid, was established when costs of solar and other BTM DG were high and deployment was low. However, rapid technological change and cost reductions over the last decade combined with a range of supporting policies, including net metering compensation rates, have led to a rapid growth in the proliferation of DG. As a result, DG technologies are at a stage, both technologically and economically, where net metering policies need to be revisited and evolved in considering how to support DG deployment in the context of the evolving electricity system.

This report explores the evolution of net metering and the changes needed to continue to support and advance BTM DG in a way that leads to a decarbonizing, equitable, and resilient electricity system. This chapter builds on the detailed consideration of economic, equity, technological, regulatory, and market factors presented in the earlier chapters. In considering the recommendations in this report, which are summarized in this chapter, it is important to bear in mind that many of the efficiency and equity issues that arise in the context of BTM DG can be addressed either by adjusting the rate designs that underlie net metering compensation or by changes directly to net metering and its variants, and by consideration and potential adoption of alternative support mechanisms. While the focus of this report has been on net metering, its variants, and alternatives, broader changes to underlying rate designs would establish a more level playing field for how changes to net metering and its variants manifest in terms of a decarbonizing, equitable, and resilient electricity system.

EVOLVING NET METERING FOR THE FUTURE ELECTRICITY SYSTEM

Net metering was instituted in the 1980s, both for its technical simplicity and customer ease of understanding, for supporting DG, especially rooftop solar, with a view to encourage private investment in renewable energy resources, among other objectives.[1] Measured against these original objectives, net metering has been quite successful.

In fact, net metering has been so successful that some states have achieved threshold levels of BTM DG that are raising tensions with the traditional approaches of planning, designing, funding, and maintaining the electricity system. These tensions must be addressed. There is a growing recognition that the value of DG and the ability to capture it for the benefit of the system and all customers depends on the timing, location, and the nature of service (e.g., energy, demand response, and resilience) that DG provides, as well as the degree to which the DG is integrated (not just interconnected) and interoperated with the distribution grid. DG's impact is also affected by the specific utility context, public policy, and rate structures and levels.

The committee envisions and supports the evolution of net metering policies in a way that addresses both the value and equity aspects of DG. DG offers potentially multiple value streams, from energy to demand response, to decarbonization, to backup-based resilience, to reduced mortality and morbidity by offsetting generation from central fossil fuel power plants, and more. It is also clear that DG introduces variability into the demand-side of the equation—on the one hand increasing the complexity of system operations, potentially raising their cost; and on the other hand, with the right control and management technologies, potentially offering options to reduce system costs. DG at scale has both benefits and costs, which arise from the same source: the flexible and distributed nature of DG. The committee believes that the future of net metering lies in creating a rate and system architecture that appropriately recognizes *both* the benefits and costs of DG—commensurately rewarding BTM DG when it provides value for the system and charging DG customers when the system delivers value to them.

[1] Quoted from Chapter 3: "According to a 1998 study by National Renewable Energy Laboratory (NREL) researchers: 'The main objective for states implementing net metering programs is to encourage private investment in renewable energy resources. Other goals include[d] stimulating local economic growth, diversifying energy resources, and improving the environment. Since then, other objectives have been added, including giving consumers options to produce their own power and manage their electricity bills, driving down the costs of renewables as the market expands, and producing power with no carbon emissions.'"

The treatment of DG must adjust with future conditions, ensuring fair compensation for customers who invest in BTM DG and reflecting its contribution to system decarbonization and resilience.

In addition to revisiting net metering policy and exploring its variants and alternatives based on the value or benefits DG can provide, a review of net metering policy offers an important opportunity to address equity concerns that should be addressed in all policymaking. From the equity lens, BTM DG supported by net metering engenders a complex landscape of positive and negative impacts. On one hand, BTM DG positively affects communities in fossil fuel–dominated supply systems because of pollutant reductions and the associated health benefits; on the other, BTM DG can fail to address or even exacerbate historical issues of access to programs and technologies for low-income communities and cost implications for net metering non-participants. Policymakers need to focus attention on equity issues to make certain that equity will be enhanced through net metering policy for BTM DG. Unleashing the true power and promise of DG for a more resilient, equitable, and decarbonized electricity system will require a more intentional and integrated approach than traditional net metering policies.

Distributional aspects of net metering are complex and are affected by the underlying context set by electricity rates. Although different rate designs directly or indirectly have provisions to limit the adverse impact of rates on low-income and disadvantaged communities, given that rates try to *balance* many different factors, maximizing equity and access is generally not the primary goal. The proliferation of new technologies, such as DG, under the aegis of older rate designs further complicates this relationship between rates and their equity impacts. New rates and net metering and its variants should be rigorously evaluated for directly addressing equity issues and adjusted, if appropriate, to address these issues before broad implementation. Furthermore, the need for balance between the economic efficiency and the distributional impacts of net metering and its variants should be recognized upfront, and parallel incentive and support mechanisms to directly improve the equity and resilience outcomes for low-income and disadvantaged communities should be put into place in tandem with both new and existing net metering policies.

PILLARS OF NET METERING REDESIGN

The evolution of net metering needs to build on three core foundational pillars:

1. It needs to recognize that the future electricity system is and will become even more interdependent among its constituent parts (including DG at the grid edge), not less, and should accordingly support the integration of those parts.

2. It needs to build on the generally accepted principles of electricity system regulation: efficiency, simplicity, stability, fairness, and revenue adequacy.
3. It needs to be informed by the electricity system context, policy objectives related to decarbonization, equity, and system resilience, among others, and be sensitive to the locational, temporal, and scale impacts of DG.

DG as Part of an Interdependent, Synergistic Evolving Electricity System

The committee believes that through better coordination, better system outcomes may be achieved. Certainly, in the medium term (5–10 years), but even more so in the long term (20 years and beyond), the committee foresees a more interdependent future for customers and utilities and other energy product and service providers, as well as distributed and grid-scale generation. This integrated system will deliver greater value for customers overall. While there may be some customers, in certain circumstances, who may choose to disconnect from the grid, such an outcome applied to the entire or even large parts of the customer base is not likely or desirable.[2] It would destabilize and disintegrate the larger system of electricity service delivery, without guaranteeing better reliability or economics for customers individually or en masse. Rather, the committee envisions a pathway that integrates DG with the rest of the electricity system and leverages valuable features of the integrated system for the benefit of all customers.

Indeed, as highlighted throughout this report, DG offers a new set of value-driven uses such as customer savings, backup power, demand response, and lower emissions. These value streams enable better electricity service experience for customers with and without DG and offer new avenues for building a cleaner, more resilient, and more equitable electricity system. That value is dependent on time and location; estimating it and operating to achieve it will be essential. To estimate and capture this value data relating to utility and system costs must be made available and accessible to customers and DG providers. Similarly, DG can lead to increased system costs due to its variability and technical characteristics (e.g., loss of system inertia, among others), adding operational challenges at the distribution level, and the need for system investment to address them. BTM DG supported by net metering can also have distribution impacts, potentially raising equity concerns, due to cost shifts from net metering to non-net metering customers, or vice versa. Where BTM DG value is less than net metering compensation, costs will shift to non-net metering customers; where value exceeds compensation, costs may be

[2] These circumstances could occur if customers can meet their own electricity needs with BTM DG, storage and possibly backup generation, at a cost that is less than they would otherwise pay the utility. Net metering compensation for exported power can offset some of the customer's total cost and can influence the decision to remain connected (and receive such compensation) or disconnect.

shifted to net metering customers. Benefits and costs of DG also depend on the scale of DG penetration.

While the exact balance of benefits and costs is system-dependent, there is value on both sides (BTM and grid-side), even with costs and uncertainties on both sides as well. For BTM DG customers, uncertainties arise from changing prices, rate structures, and reliability and resilience of supply, whether self-generated or grid supplied. For grid operators and utilities, uncertainties come from variable demand, hard-to-predict operational conditions, changing customer and policymaker expectations, and evolving regulatory and utility business models. All of these are affecting utilities' operations, conventional revenue collection, and system and financial planning. DG evolution challenges the top-down, conventional model of the distribution grid, where decisions of operators may be at odds with customer behavior. Given the continued adoption and deployment of BTM DG, that is also increasingly combined with storage, higher levels of variability in demand are a reality. Utilities need to continue to plan for, invest in, and operate the system to meet increasingly variable customer needs and wants from the grid. Utilities also need to leverage DG attributes for building a more resilient electricity system.

Redesigned Rate Structures and Net Metering Need to Be Consistent with the Basic Principles of Electricity Rate Design

New rate structures and net metering need to build on the generally accepted regulatory principles of efficiency, simplicity, stability, fairness, and revenue adequacy. Achieving the right balance is a challenging and changing goal, yet one that makes explicit the multiple higher-level objectives that need to be balanced. Acceptable rate designs have always required tradeoffs or balancing among ratemaking principles and objectives that have been weighted differently in different locations and at different times. The committee expects that to continue as part of net metering policy evolution. In fact, because of the opportunities and challenges associated with DG at varying levels of deployment, the contours of those tradeoffs are expanding. At the heart of this new balancing act is the ability to integrate DG more completely into the electricity system for the benefit of the system and all customers while enabling private choice and flexibility. Where the customer maintains full control over the timing and duration of power drawn from or injected into the grid, the DG is operationally uncontrollable from the grid's viewpoint and is of less value to the grid and other customers. Where DG resources offer control to the utility—either directly or indirectly through price signals or contracts—to help maximize value for the grid, they provide more value and thus merit greater compensation.

BTM DG offers multiple value streams (e.g., lower costs, lower pollution, and higher resilience) for both the customer and the grid. At higher levels of penetration, it also introduces new challenges for the distribution system related to demand variability and power flows. Capturing these value streams and reflecting the values in compensation through net metering or its variants while addressing the emerging challenges will entail sustained investments in the distribution grid in order to integrate—not just interconnect—BTM DG (and other DER) into the grid. When those investments are best made by the utility, ratemaking should enable the utility to earn a reasonable return on them. However, the committee also recognizes that some investments for integrating BTM DG into the grid, for example through demand aggregation or virtual power plant mechanisms, could also be made by non-utility market participants.[3]

Thus, coordinating the evolution of net metering with broader electricity system changes will lead to a vastly superior outcome than addressing net metering without due consideration of its context. Solutions that account for system- and context-specific nuances will be needed and can be achieved through consultation and coordination. Coordination mechanisms and processes will be key, placing emphasis on distributed decision making under uncertainty and distributed control systems. Moreover, regardless of the complexities of the technical systems, the rate structures and net metering compensation for BTM DG should be straightforward and easy for customers and other stakeholders to understand.

System Context and History Should Inform Net Metering Redesign

Good net metering redesign will need to be situated in the respective system context, policy objectives, and history. The committee does not believe that there is just one net metering model that is categorically superior for all systems and contexts. Important elements of context could include DG resource and potential, relevant policies (carbon and pollution reduction, resilience, equity, and others), current DG deployment levels, and expected system stressors (due to climate change and other factors). Important elements of system history could include current electricity infrastructure and near- and medium-term revenue requirements due to historical grid investments. In fact, even for the same system or utility, elements of net metering design that appropriately balances different tradeoffs will likely vary over time and as DG deployment changes. The benefits and costs of BTM DG, supported by net metering, at lower levels of deployment can

[3] It will be important to ensure that non-utility participants do not face barriers to making these investments and providing these services (e.g., standardized interconnection and interoperability requirements would reduce those barriers).

be quite different from when DG deployment is greater and increasing. As the costs of BTM DG systems and storage continue to decline, the potential for greater numbers of customers to adopt these technologies must be considered. While the core lessons and insights of net metering policy evolution and redesign will be applicable across different systems, each system will nonetheless need to grapple with its unique set of context, history, and aspirations.

"BEYOND DG" APPROACH NEEDED FOR NET METERING REDESIGN

The proliferation of BTM DG is happening within a larger set of changes in the electricity system and interconnected systems (e.g., telecommunications). Maximizing the benefit for all customers through the evolution of net metering will entail successfully engaging with these broader systems and issues and co-evolving with them.

One of the major underlying trends has been the digitization of the electricity system over the past few decades.[4] This is not unique to the electricity industry, rather it is part of what has been broadly dubbed as *Industry 4.0*—the coming together of objects (e.g., machines) in the physical infrastructure, their virtual representations, and services enabled through the use of data and analytics by leveraging that physical-virtual combination.[5] Digitization within the electricity distribution system, including an intertwined array of smart hardware (e.g., meters and sensors) and software (e.g., DG performance dashboards and automated fault detection), has played a central role in making BTM DG and DER integration possible.[6,7] Expected to further deepen in the future, digitization will be central to the emergence of coordinated, hierarchical multi-agent systems helping better integrate DER into the grid, further enabling the robustness of microgrids, energy hubs, intelligent energy management, and system-level optimization.[8]

The committee notes that, among others, the following areas are especially worthy of continued attention in the context of net metering policy evolution:

---

[4] IEA (International Energy Agency). 2017. *Digitalisation and Energy*. Paris: IEA. https://www.iea.org/reports/digitalisation-and-energy.

[5] Drath, R., and A. Horch. 2014. "Industrie 4.0: Hit or Hype? [Industry Forum]." *IEEE Industrial Electronics Magazine* 8(2):56–58.

[6] Bañales, S. 2020. "The Enabling Impact of Digital Technologies on Distributed Energy Resources Integration." *Journal of Renewable and Sustainable Energy* 12(4):045301.

[7] Venugopalan, S., and V. Rai. 2015. "Topic Based Classification and Pattern Identification in Patents." *Technological Forecasting and Social Change* 94:236.

[8] Howell, S., Y. Rezgui, J.L. Hippolyte, B. Jayan, and H. Li. 2017. "Towards the Next Generation of Smart Grids: Semantic and Holonic Multi-Agent Management of Distributed Energy Resources." *Renewable and Sustainable Energy Reviews* 77:193–214.

- An expanding array of economic technology solutions including photovoltaic (PV) + storage + generator, combined heat and power (CHP), and Plug-and-Play (BTM DG) will be available to customers
- The rapid evolution of DG, storage, and generation options, along with microgrid technologies, that may enable system resilience through distributed capabilities (in addition to any private resilience value afforded to adopters of these technologies)
- Increasing electrification of transportation, homes, and buildings, with accompanying consumer demand response capabilities and technologies
- The emergence of new customer groups and assets (e.g., through community choice aggregation and demand response aggregation)
- The need to reinforce and expand system communication technologies and infrastructure
- The increasing challenges to cybersecurity on a power grid with increasing numbers of distributed and interconnected devices

Exploration and implementation of good net metering policy redesigns, especially in electricity systems with relatively high or soon-to-be high DG penetration, will need to maintain a holistic look at this "integrated DG frontier." Focusing on the singular question of benefit and cost for BTM customers vis-à-vis the grid operator and the utility will miss the larger set of intertwined questions that are at play.

ELECTRICITY RATES AND NET METERING POLICIES ARE NOT A SUBSTITUTE FOR ADDRESSING SOCIETAL NEEDS MORE BROADLY

The nature and extent to which different externalities associated with the production and supply of electricity may be included in rates are governed by relevant state and federal legal provisions applicable within the jurisdiction in which a utility operates. At present, most jurisdictions in the United States do not require relevant externalities (e.g., GHG emissions) to be priced into utility rates. This reality poses a fundamental constraint for ratemaking processes, continuing the gap between pricing at social marginal cost, based on economic theory, and electricity rates, which balance economic efficiency and other objectives. In this sense, the regulatory ratemaking framework and processes establish the bounds for net metering (and other policies). Updating that framework through policymaking that enables the valuation and internalization (i.e., inclusion in prices) of the relevant externalities will be important for leveling the playing field for various grid resources and appropriately valuing them through rates. However, there will be limits to this. Consequently, it is vital to consider and support non-electricity system goals (such as economy-side carbon reduction, technology innovation, and low-income support) outside the rate design arena.

SUMMARY

Thanks to rapid technological advancements and cost reductions, DER generally, and BTM DG (especially distributed solar) more specifically are playing a more central and active role in the electricity system than was thought possible a couple of decades ago. Net metering has successfully supported the growth of BTM DG during this period. Moving forward, to enable the integration of BTM DG into the electricity system to maximize benefits for all customers, traditional net metering needs to be revised to align value streams, costs, and compensation. This would be the first key priority of regulatory action in this area. Second, it would be critical to ensure that revisions to net metering also improve equity, an outcome that cannot be taken for granted, rather one that must be achieved through a continual, intentional approach. Achieving these dual objectives of greater economic efficiency and equity would require careful attention to and balancing of contextually informed factors that often vary across different electricity systems. Fully integrating BTM DG into the electricity system portends a more resilient electricity system, but achieving this goal would entail systematic, coordinated, and sustained investments across distribution systems. Policies and regulations must support the creation of a framework and market conditions that enable utility and, as applicable, non-utility participants to invest in the distribution system for integrating BTM DG into the grid, with an opportunity to earn a reasonable rate of return on their investment. Finally, not everything can or should be achieved through electricity rates. Broader social goals may be best pursued through non-rate approaches, and additional supporting mechanisms that go beyond ratemaking will likely be necessary for attaining certain key objectives, especially equity. With a more open, transparent, deliberate, and intentional approach that leverages the locational, temporal, and contextual value streams of BTM DG, an evolved net metering policy could enable an electricity system that is more sustainable, equitable, and resilient.

RECOMMENDATIONS

Trends

Recommendation 3-1: Utility regulators and decision makers of publicly owned utilities should work with stakeholders to explore and implement solutions to technical, legal, proprietary, or privacy concerns about the collection and release of

information about the performance of behind-the-meter technologies and their interactions with the grid.

Recommendation 3-2: Data about participation in net metering and alternative and supplemental DG policies should be collected and reported so that policy-makers can monitor how policies are interacting with energy market conditions in wholesale and retail, as well as regulated and unregulated, energy markets.

Economics

Recommendation 4-1: Regulators should strive to develop retail rate structures—for both DG and non-DG customers—with usage-based energy prices that correspond as closely as possible to the social marginal cost of producing and delivering electricity, inclusive of time- and location-specific environmental or other externality costs (insofar as they are not already internalized), while recognizing other competing rate-design objectives. Under this condition, net metering will generally result in economically efficient levels of investment in DG technologies, albeit with some rate impacts to non-participating customers.

Recommendation 4-2: In the absence of economically efficient rate structures for all customers, the alternative solution may be to implement changes to the net metering mechanism—either buy-all and sell-all or net billing tariffs—for DG customers, with DG compensation levels set at or near the social marginal cost of electricity production and delivery.

Recommendation 4-3: When evaluating the economic implications of current or proposed changes to net metering rules, decision makers should consider specific conditions in their jurisdiction (e.g., the generation mix and costs, DG penetration levels, clean energy policies, etc.), while also anticipating how those conditions may evolve over the lifetime of DG resources. Environmental benefits should be carefully weighed alongside any rate impacts to non-participating customers.

Recommendation 4-4: In some cases, additional compensation for DG customers may be warranted above and beyond what can be provided through net metering or other DG tariffs. In these circumstances, any additional incentives for DG customers should be provided through transparent incentives designed to achieve a

specified outcome (e.g., to encourage adoption by low-income customers, to relieve constraints in specific localized regions of the grid, to meet legislative DG targets, etc.) and, if appropriate, funded from sources outside of electricity rates.

Recommendation 4-5: In determining DG compensation, policymakers should explicitly account for the environmental and other externalities that DG reduces when it displaces fossil fuel generation. To fairly compensate DG for this clean energy attribute, policymakers should consider whether to do so based on the monetary value of the environmental externalities that the DG is likely to abate or the cost to achieve the same reduction in environmental harms through alternative technologies or policies.

Recommendation 4-6: Utilities should work with regulators and stakeholders to identify the types and formats of information (such as heat maps and avoided-cost calculations) that the utility should provide publicly about their distribution systems and their provision of electricity service so that suppliers and potential customers of behind-the-meter generation can see the benefits and costs that behind-the-meter technologies can provide to the local electric system. Regulators and decision makers of publicly owned utilities should ensure that such information is provided to the public in easily understandable and accessible ways.

Recommendation 4-7: To the extent possible, BTM DG compensation rates should be location- and time-based, tied to available distribution capacity, to indicate where that DG would have the greatest value (e.g., in relieving or avoiding distribution constraints and/or displacing local generation with high emissions). Considering the locational value of solar can have particular value to low-income customers who often live in areas with emitting generation and poorer air quality.

Equity

Recommendation 5-1: Rates should be designed consistent with updated ratemaking principles, with particular attention to the equity impacts for customers least able to afford them. To help accomplish this, utilities and policymakers should ensure that information about utility rates is easily available to all customers (e.g., not just through postings on websites), and that all customers have a voice and can participate in the design of rates.

Recommendation 5-2: Decision makers on electricity rates should consider both the impacts of the distribution of benefits and costs, as well as total benefits and costs when designing net metering policies, and ensure that adequate data are collected and made publicly available to do so. These benefits and costs should include and balance, among other things: public health impacts, job impacts, land use impacts, and the future options that will be enabled or precluded.

Recommendation 5-3: Fair ratemaking should allow for affected consumers to have a meaningful voice in the process (so as to ensure procedural equity) and take into consideration (among other things) the potential impacts of rates on various customer segments (from an economic efficiency and equity point of view), and on customers' decisions about adopting BTM technologies or load or grid defection in the context of technology trends. Regulators and decision makers of publicly owned utilities should ensure that such information is provided to the public and is easily understandable and accessible.

Recommendation 5-4: Investments to reverse historic structural inequities, such as grid upgrades in historically disinvested communities to increase hosting capacity, should be considered to enable broader adoption of BTM DG and participation in net metering.

Recommendation 5-5: Policymakers should consider expanding community and government-funded programs, such as low- or no-interest financing, to expand access to BTM DG by low- and medium-income customers and their participation in net metering. In addition, incentives that focus on income-qualified customers and renters can help reduce inequities due to lack of access.

Technology

Recommendation 6-1: There need to be direct investments in the distribution system to integrate increasing amounts of BTM DG such as rooftop solar, as well as other DER including smart buildings management systems, electric vehicles, and charging infrastructure, to ensure the continued safe and reliable operation of the grid and provision of grid services. Investments will also be required to improve grid visibility to suitably site and operate BTM DG, as well as provide efficient price signals, such that the DG can provide system benefits, particularly local and grid resilience when normal service is disrupted. These investments in the distribution grid have to occur simultaneously with BTM DG deployment.

Recommendation 6-2: In order to make the best use of DG and DER, utilities must make investments to integrate these technologies, increase their visibility, manage them (either directly or indirectly through price signals), and reduce barriers to their management by customers and DG providers. With either utility or non-utility control, or intelligent management, DG can provide greater value. The corresponding compensation through net metering and its variants could reflect this higher value. Overall, the democratization of the grid, with increased public and private partnerships, can be very valuable.

Recommendation 6-3: Investments in distribution system technologies aimed toward integration of DG and DER must be accompanied by revisions in policies and state and federal utility regulations to facilitate cost recovery of these investments.

Regulatory and Policy

Recommendation 7-1: Decision makers about electric utility rates—including state legislators, utility regulators, and governing boards of publicly owned electric utilities—should take into account that DG technology costs and market maturity are at a stage both technically and economically, where traditional net metering policies to support the deployment of DGs need to be assessed and revisited. This recommendation applies both to instances where a utility operates in a state that previously adopted net metering and regulators are considering variants to it, as well as in parts of the country that have not yet adopted net metering, and seek to advance BTM technologies, and have the option to leapfrog beyond net metering and adopt other ratemaking variants.

Recommendation 7-2: Decision makers should rely on important, traditional ratemaking principles as updated to reflect the application of new technologies and service offerings. These updated principles include cost-causation, rate simplicity, fairness, revenue adequacy, a simple customer experience even if underlying rate structures become significantly more sophisticated, and compensating resources based on their value.

Recommendation 7-3: Decision makers should design compensation approaches for the export of power from BTM generation according to principles that are consistent with how the utility values other sources of power that offer comparable energy, capacity, and other grid services to the system (which may vary by time and location). External impacts (such as pollution) from some sources of power are

unpriced, and in many jurisdictions there are constraints on the ability of regulators to reflect externalities in utility planning and/or ratemaking. Sound economic principles would support the consideration of such externalities in utility regulation; policymakers should consider how to address such impacts in utility and other energy policies.

Recommendation 7-4: Given the economic and equity challenges associated with using net metering—and with financial incentives and programs recovered in electricity rates more generally—to promote investment in and deployment of DG technologies, policymakers should also consider and where appropriate use other policy instruments, such as tax incentives, building codes, attractively priced loans (or even grants) to low-income households, and other complementary policies.

Bibliography

Aflaki, S., and A. Masini. 2020. "Optimizing Feed-In Tariffs to Boost Renewable Energy Production." Jouy en Josas, Paris: HEC Paris. https://www.hec.edu/en/knowledge/articles/optimizing-feed-tariffs-boost-renewable-energy-production.

Allen, M. 2003. "Liability for Climate Change." *Nature* 421(6926):891–892.

Allen, N., A. Clair, M. McNaul, R. Shelton, and J. Williams. 2021. *PURPA: Title II Compliance Manual 2.0*. Thompson Coburn LLP. https://pubs.naruc.org/pub/47AD30DC-1866-DAAC-99FB-975A60906D6B.

Altunay, M., A. Bergek, and A. Palm. 2021. "Solar Business Model Adoption by Energy Incumbents: The Importance of Strategic Fit." *Environmental Innovation and Societal Transitions* 40:501–520.

American Council for an Energy-Efficient Economy. 2020. "Report: Low-Income Households, Communities of Color Face High 'Energy Burden' Entering Recession." https://www.aceee.org/press-release/2020/09/report-low-income-households-communities-color-face-high-energy-burden.

Anderson, W., V. White, and A. Finney. 2012. "Coping with Low Incomes and Cold Homes." *Energy Policy* 49:40–52.

Arizona Corporation Commission. 2017. "Decision No. 75859." Docket No. E-00000J-14-0023, January 3, 2017. https://docket.images.azcc.gov/0000176114.pdf?i=1614371719161.

Audia, C., F. Berkhout, G. Owusu, Z. Quayyum, and S. Agyei-Mensah. 2021. "Loops and Building Blocks: A Knowledge Co-Production Framework for Equitable Urban Health." *Journal of Urban Health* 98:394–403.

Bai, J., D. Shuai, and E. Tian. 2022. "Solar PV Battery Installations in Honolulu: 2021 Update." Research and Economic Analysis Division, Department of Business, Economic Development and Tourism, State of Hawaii. https://files.hawaii.gov/dbedt/economic/data_reports/reports-studies/Solar_PV_Battery_Installation_2022May.pdf.

Baja, K. 2018. "Resilience Hubs: Shifting Power to Communities and Increasing Community Capacity." Urban Sustainability Directors Network (USDN). https://www.usdn.org/uploads/cms/documents/usdn_resiliencehubs_2018.pdf.

Bañales, S. 2020. "The Enabling Impact of Digital Technologies on Distributed Energy Resources Integration." *Journal of Renewable and Sustainable Energy* 12(4):045301.

Barbose, G. 2017. *Putting the Potential Rate Impacts of Distributed Solar into Context*. Berkeley, CA: Lawrence Berkeley National Laboratory. https://emp.lbl.gov/publications/putting-potential-rate-impacts.

Barbose, G. 2021. *U.S. Renewables Portfolio Standards 2021 Status Update: Early Release*. Berkeley, CA: Lawrence Berkeley National Laboratory. https://emp.lbl.gov/publications/us-renewables-portfolio-standards-3.

Barbose, G., and N. Darghouth. 2019. *Tracking the Sun: Pricing and Design Trends for Distributed Photovoltaic Systems in the United States*. Berkeley, CA: Lawrence Berkeley National Laboratory. https://emp.lbl.gov/sites/default/files/tracking_the_sun_2019_report.pdf.

Barbose, G., N. Darghouth, E. O'Shaughnessy, and S. Forrester. 2022. *Tracking the Sun: Pricing and Design Trends for Distributed Photovoltaic Systems in the United States*. Berkeley, CA: Lawrence Berkeley National Laboratory. https://emp.lbl.gov/tracking-the-sun.

Barbose, G.L., S. Elmallah, and W. Gorman. 2021a. *Behind the Meter Solar + Storage: Market Data and Trends*. Berkeley, CA: Lawrence Berkeley National Laboratory. https://emp.lbl.gov/publications/behind-meter-solarstorage-market-data.

Barbose, G.L., S. Forrester, E. O'Shaughnessy, and N.R. Darghouth. 2021b. *Residential Solar-Adopter Income and Demographic Trends: 2021 Update*. Berkeley, CA: Lawrence Berkeley National Laboratory. https://emp.lbl.gov/publications/residential-solar-adopter-income-and.

Barbose, G., and A.J. Satchwell. 2020. "Benefits and Costs of a Utility-Ownership Business Model for Residential Rooftop Solar Photovoltaics." *Nature Energy* 5(10):750–758.

Barnea, G., C. Hagemann, and S. Wurster. 2022. "Policy Instruments Matter: Support Schemes for Renewable Energy Capacity in Worldwide Comparison." *Energy Policy* 168.

Barnes, K., B. Johnson, and R. Nickelson. 2004. *Review of Supervisory Control and Data Acquisition (SCADA) Systems*. Idaho Falls, ID: Idaho National Engineering and Environmental Laboratory. https://inldigitallibrary.inl.gov/sites/sti/sti/3310858.pdf.

Basu, R. 2009. "High Ambient Temperature and Mortality: A Review of Epidemiologic Studies from 2001 to 2008." *Environmental Health* 8:40.

BECC (Behavior, Energy and Climate Change). 2022. "About BECC." https://beccconference.org/about-becc.

Bixler, P., J.A. Belaire, K.M. Faust, M. Scoggins, and A. González. 2022. "Exploring the Connection Between Transdisciplinary Co-Production and Urban Sustainability Solutions: A Case Study at an Urban Stream Management Symposium." *Urban Ecosystems* 25:1207–1216.

Blackburn, G., C. Magee, and V. Rai. 2014. "Solar Valuation and the Modern Utility's Expansion into Distributed Generation." *The Electricity Journal* 27(1):18–32.

Bollinger, B., and K. Gillingham. 2012. "Peer Effects in the Diffusion of Solar Photovoltaic Panels." *Marketing Science* 31(6):900–912.

Bonbright, J.C., A.L. Danielsen, and D.R. Kamerschen. 1988. *Principles of Public Utility Rates*. 2nd ed. Dumfries, VA: Public Utility Reports, Inc.

Borenstein, S., M. Fowlie, and J. Sallee. 2021. *Designing Electricity Rates for an Equitable Energy Transition*. Berkeley, CA: Energy Institute at HAAS. https://haas.berkeley.edu/wp-content/uploads/WP314.pdf.

Brockway, A.M., J. Conde, and D. Callaway. 2021. "Inequitable Access to Distributed Energy Resources Due to Grid Infrastructure Limits in California." *Nature Energy* 6:892–903.

Brown, D.P. 2022. "Socioeconomic and Demographic Disparities in Residential Battery Storage Adoption: Evidence from California." *Energy Policy* 164:112877.

Brown, M.A. 2022. "Testimony Before the Georgia Public Service Commission on Georgia Power's 2022 Rate Case." Document Filing 192146. Georgia Public Service Commission. https://psc.ga.gov/search/facts-document/?documentId=192146.

Brown, M.A., and M. Ahmadi. 2019. "Would a Green New Deal Add or Kill Jobs?" *Scientific American*. https://www.scientificamerican.com/article/would-a-green-new-deal-add-or-kill-jobs1.

Brown, M.A., J. Hubbs, X.V. Gu, and M.-K. Cha. 2021. "Rooftop Solar for All: Closing the Gap Between the Technically Possible and the Achievable Potentials." *Energy Research and Social Science* 80(October):102203.

Brown, M.A., A. Soni, M.V. Lapsa, and K.A. Southworth. 2020a. *Low-Income Energy Affordability: Conclusions from a Literature Review*. Oak Ridge, TN: Oak Ridge National Laboratory.

Brown, M.A., A. Soni, M.V. Lapsa, K.A. Southworth, and M. Cox. 2020b. "High Energy Burden and Low-Income Energy Affordability: Conclusions from a Literature Review." *Progress in Energy* 2(4):1–35.

Burger, S., I. Schneider, A. Botterud, and I. Pérez-Arriaga. 2018. *Fair, Equitable, and Efficient Tariffs in the Presence of Distributed Energy Resources*. Cambridge, MA: Massachusetts Institute of Technology Center for Energy and Environmental Policy Research.

Burton, R., and X. Kaifeng. 2022. "Community Solar Resource Database." Golden, CO: National Renewable Energy Laboratory. https://data.nrel.gov/submissions/201.

California Energy Commission. 2022. "Solar PV Systems and Solar Ready." https://www.energy.ca.gov/programs-and-topics/programs/building-energy-efficiency-standards/online-resource-center/solar.

California Energy Commission. 2023. "SB 100 Joint Agency Report." https://www.energy.ca.gov/sb100.

California Legislative Information. 1995. "An Act to Add Section 2827 to the Public Utilities Code, Relating to Public Utilities." State of California Senate Bill No. 656. http://www.leginfo.ca.gov/pub/95-96/bill/sen/sb_0651-0700/sb_656_bill_950804_chaptered.html.

Campbell, H., and D. Vanderhoven. 2016. *Knowledge That Matters: Realising the Potential of Co-Production*. Manchester, UK: N8 Research Partnership.

Carley, S., and D.M. Konisky. 2020. "The Justice and Equity Implications of the Clean Energy Transition." *Nature Energy* 5:569–577.

Carvallo, J.P., F.C. Hsu, Z. Shah, and J. Taneja. 2021. *Frozen Out in Texas: Blackouts and Inequity*. New York: The Rockefeller Foundation.

Casey, J.A., J.G. Su, L.R. Henneman, C. Zigler, A.M. Neophytou, R. Catalano, R. Gondalia, Y-T Chen, L. Kaye, S.S. Moyer, V. Combs, G. Simrall, T. Smith, J. Sublett, and M.A. Barrett. 2020. "Coal-Fired Power Plant Closures and Retrofits Reduce Asthma Morbidity in the Local Population." *Nature Energy* 5(5):365–366.

Castanda, M., M. Jimenez, S. Zapata, C.J. Franco, and I. Dyner. 2017. "Myths and Facts of the Utility Death Spiral." *Energy Policy* 110:105–116.

Cazalet, E.G, M. Kohanim, and O. Hasidim. 2020. *Complete and Low-Cost Retail Automated Transactive Energy System (RATES)*. CEC 500-2020-038. Prepared for the California Energy Commission. Encino, CA: Universal Devices, Inc. https://www.energy.ca.gov/sites/default/files/2021-05/CEC-500-2020-038.pdf.

Center for Climate and Energy Solutions. 2023. "U.S. State Carbon Pricing Policies." https://www.c2es.org/document/us-state-carbon-pricing-policies.

Center for Sustainable Systems. 2021. *U.S. Grid Energy Storage Factsheet*. No. CSS15-17. Ann Arbor, MI: University of Michigan. https://css.umich.edu/publications/factsheets/energy/us-grid-energy-storage-factsheet.

CESA (Clean Energy States Alliance). n.d. "Table of 100 Percent Clean Energy States." https://www.cesa.org/projects/100-clean-energy-collaborative/guide/table-of-100-clean-energy-states.

Chan, G., J. Heeter, and K. Xu. 2022. "Sharing the Sun Community Solar Project Data (December 2021)." Golden, CO: National Renewable Energy Laboratory. https://data.nrel.gov/submissions/185.

Chew, B., E.H. Myers, T. Adolf, and E. Thomas. 2018. "Non-Wires Alternatives, Case Studies from Leading U.S. Projects." Smart Electric Power Alliance, Peak Load Management Alliance, and E4TheFuture. https://e4thefuture.org/wp-content/uploads/2018/11/2018-Non-Wires-Alternatives-Report_FINAL.pdf.

Clean Energy Group. 2023. "Resilient Power Project." https://www.cleanegroup.org/publication-project/resilient-power-project.

Cleary, K., and K. Palmer. 2022. *US Electricity Markets 101*. Washington, DC: Resources for the Future. https://www.rff.org/publications/explainers/us-electricity-markets-101.

Cleveland, M., L. Dunning, and J. Heibel. 2019. "State Policies for Utility Investment in Energy Efficiency." National Conference of State Legislatures. https://www.ncsl.org/Portals/1/Documents/energy/Utility_Incentives_4_2019_33375.pdf?ver=2019-04-04-154310-703.

Climate Xchange. 2022a. "State Climate Policy Tracker." https://climate-xchange.org/network/map.

Climate Xchange. 2022b. "State Climate Policy Tracker (Spreadsheet), Version 1.3." https://docs.google.com/spreadsheets/d/1t912_uR-x8DeMTDNCm6y9nb6QLiiLDAdWNciteu1rOU/edit#gid=951166054.

Colorado State University, Center for the New Energy Economy, and The Nature Conservancy. "Methodology Behind the State Policy Opportunity Tracker (SPOT) for Clean Energy." https://spotforcleanenergy.org/about.

Colton, R.D. 2011. "Home Energy Affordability in New York: The Affordability Gap (2008–2010)." Prepared for New York State Energy Research Development Authority. https://www.nyserda.ny.gov/-/media/Project/Nyserda/Files/EDPPP/LIFE/Resources/2008-2010-affordability-gap.pdf.

Converge Strategies, LLC. 2019. "The Value of Resilience for Distributed Energy Resources: An Overview of Current Analytical Practices." Prepared for the National Association of Regulatory Utility Commissioners. https://pubs.naruc.org/pub/531AD059-9CC0-BAF6-127B-99BCB5F02198.

Cook, E.M., M. Berbés-Blázquez, L.M. Mannetti, N.B. Grimm, D.M. Iwaniec, and T.A. Muñoz-Erickson. 2021. "Setting the Stage for Co-Production," pp. 99–111 in *Resilient Urban Futures, The Urban Book Series*. Cham: Springer.

Cook, J.J., and M. Shah. 2018. *Reducing Energy Burden with Solar: Colorado's Strategy and Roadmap for States*. NREL/TP-6A20-70965. Golden, CO: National Renewable Energy Laboratory.

Cook, J.T., P.H. Casey, R. Rose-Jacobs, M.M. Black, M. Chilton, S. Ettinger de Cuba, D. Appugliese, S. Coleman, T. Heeren, C. Berkowitz, and D.B. Cutts. 2008. "A Brief Indicator of Household Energy Security: Associations with Food Security, Child Health, and Child Development in US Infants and Toddlers." *Pediatrics* 122(4):e867–e875.

Council on Environmental Quality. 2022. "Climate and Economic Justice Screening Tool." https://screeningtool.geoplatform.gov/en/#3/33.47/-97.5.

CPUC (California Public Utilities Commission). 2021a. "Distribution Planning." https://www.cpuc.ca.gov/industries-and-topics/electrical-energy/infrastructure/distribution-planning.

CPUC. 2021b. "IDER Cost-Effectiveness, Avoided Cost Calculator." https://www.cpuc.ca.gov/industries-and-topics/electrical-energy/demand-side-management/energy-efficiency/idsm.

CPUC. 2021c. "Rule 21 Interconnection." https://www.cpuc.ca.gov/rule21.

CPUC. 2021d. "Self-Generation Incentive Program (SGIP)." https://www.cpuc.ca.gov/industries-and-topics/electrical-energy/demand-side-management/self-generation-incentive-program.

CPUC. 2021e. "The Solar on Multifamily Affordable Housing (SOMAH) Program." https://www.cpuc.ca.gov/industries-and-topics/electrical-energy/demand-side-management/somah.

CPUC. 2022a. "CSI Single-Family Affordable Solar Homes (SASH) Program." https://www.cpuc.ca.gov/industries-and-topics/electrical-energy/demand-side-management/california-solar-initiative/csi-single-family-affordable-solar-homes-program.

CPUC. 2022b. "Decision on Transportation Electrification Policy and Investment." Proposed Decision, Rulemaking 18-12-006. https://docs.cpuc.ca.gov/PublishedDocs/Efile/G000/M497/K622/497622010.PDF.

CPUC. 2022c. "Order Instituting Rulemaking to Revisit Net Energy Metering Tariffs Pursuant to Decision 16-01-044, and to Address Other Issues Related to Net Energy Metering." Decision proposed to CPUC. Rulemaking 20-08-020. https://docs.cpuc.ca.gov/PublishedDocs/Published/G000/M499/K921/499921246.PDF.

CPUC. 2022d. "Resiliency and Microgrids." https://www.cpuc.ca.gov/resiliencyandmicrogrids.

Cybersecurity & Infrastructure Security Agency. 2021. "Cyber-Attack Against Ukrainian Critical Infrastructure." https://www.cisa.gov/news-events/ics-alerts/ir-alert-h-16-056-01.

Darghouth, N.R., G. Barbose, J. Zuboy, P.J. Gagnon, A.D. Mills, and L. Bird. 2020. "Demand Charge Savings from Solar PV and Energy Storage." *Energy Policy* 146(November):111766.

Darghouth, N.R., E. O'Shaughnessy, S. Forrester, and G. Barbose. 2022. "Characterizing Local Rooftop Solar Adoption Inequity in the US." *Environmental Research Letters* 17(3):034028.

Davies, L.L., and S. Carley. 2017. "Emerging Shadows in National Solar Policy? Nevada's Net Metering Transition in Context." *The Electricity Journal* 30(1).33–42.

Davis, L. 2020. "Putting Solar in the All the Wrong Places." *Energy Institute Blog*. Berkley, CA: Energy Institute at HAAS. https://energyathaas.wordpress.com/2020/02/03/putting-solar-in-all-the-wrong-places.

Day, R., G. Walker, and N. Simcock. 2016. "Conceptualising Energy Use and Energy Poverty Using a Capabilities Framework." *Energy Policy* 93:255–264.

del Rio, P., and P. Mir-Artiques. 2014. *A Cautionary Tale: Spain's Solar PV Investment Bubble*. Winnipeg, Manitoba, Canada: International Institute for Sustainable Development. https://www.iisd.org/gsi/sites/default/files/rens_ct_spain.pdf.

Denholm, P., M. O'Connell, G. Brinkman, and J. Jorgenson. 2015. *Overgeneration from Solar Energy in California: A Field Guide to the Duck Chart*. Golden, CO: National Renewable Energy Laboratory. https://www.nrel.gov/docs/fy16osti/65453.pdf.

DeShazo, J.R., and R. Matulka. 2009. "Best Practices for Implementing a Feed-in Tariff Program." Los Angeles Business Council and UCLA Luskin Center for Innovation. https://innovation.luskin.ucla.edu/wp-content/uploads/2019/03/Best_Practices_for_Implementing_a_Feed-in_Tariff_Program.pdf.

DOE (Department of Energy). 2007. *The Potential Benefits of Distributed Generation and the Rate-Related Issues That May Impede Its Expansion: Report Pursuant to Section 1817 of the Energy Policy Act of 2005*. Washington, DC: Department of Energy. https://www.energy.gov/sites/prod/files/oeprod/DocumentsandMedia/1817_Report_-final.pdf.

DOE. 2021a. "DOE Sets 2025 Community Solar Target to Power 5 Million Homes." https://www.energy.gov/articles/doe-sets-2025-community-solar-target-power-5-million-homes.

DOE. 2021b. "Solar Futures Study." https://www.energy.gov/eere/solar/solar-futures-study.

DOE. 2022. "Grid Modernization Initiative." https://www.energy.gov/gmi/grid-modernization-initiative.

DOE. 2023. "Community Solar Market Trends." https://www.energy.gov/communitysolar/community-solar-market-trends.

DOE. n.d. "Clean Energy for Low Income Communities: Single Family Home Retrofits." https://betterbuildingssolutioncenter.energy.gov/CELICA-Toolkit/single-family-home-retrofits.

DOE Global Energy Storage Database. 2021. "Search Projects." https://sandia.gov/ess-ssl/gesdb/public/projects.html.

DOE Grid Deployment Office. 2022. "National Transmission Planning Study." https://www.energy.gov/gdo/national-transmission-planning-study.

DOE Grid Modernization Initiative. 2020. "Grid Modernization Updated GMI Strategy 2020." https://www.energy.gov/sites/prod/files/2021/02/f82/GMI_Strategy_FINAL%20as%20of%201.20.21.pdf.

DOE Office of Electricity. 2019. "North American Energy Resilience Model." https://www.energy.gov/sites/prod/files/2019/07/f65/NAERM_Report_public_version_072219_508.pdf.

DOE Office of Electricity. n.d. "Recovery Act: Time Based Rate Programs." https://www.smartgrid.gov/recovery_act/time_based_rate_programs.

DOE Office of Energy Efficiency & Renewable Energy. n.d. "Solar and Resilience Basics." https://www.energy.gov/eere/solar/solar-and-resilience-basics.

Dolšak, N., and A. Prakash. 2022. "Three Faces of Climate Justice." *Annual Review of Political Science* 25:283–301.

Drath, R., and A. Horch. 2014. "Industrie 4.0: Hit or hype? [Industry Forum]." *IEEE Industrial Electronics Magazine* 8(2):56–58.

Drehobl, A., L. Ross, and R. Ayala. 2020. *How High Are Household Energy Burdens? An Assessment of National and Metropolitan Energy Burden across the United States*. Washington, DC: American Council for an Energy-Efficient Economy.

E3. 2022. "2022 Distributed Energy Resources Avoided Cost Calculator Documentation." https://www.cpuc.ca.gov/-/media/cpuc-website/divisions/energy-division/documents/demand-side-management/acc-models-latest-version/2022-acc-documentation-v1a.pdf.

E9 Insight. 2021a. "Pathways to Changing the PUC Mandate: A Regulatory Review." Prepared on behalf of the Institute for Market Transformation. https://drive.google.com/file/d/1c244_ryG4C15vnG1tuGu_1m8wgn9DwnK/view.

E9 Insight. 2021b. "PUC Mandate Database—Final." Prepared on behalf of the Institute for Market Transformation. https://docs.google.com/spreadsheets/d/1acwblw_Q67d_bpdKxFQzJbsvP19WGj59/edit#gid=286574870.

E-ISAC (Electricity Information Share and Analysis Center). 2016. *Analysis of the Cyber Attack on the Ukrainian Power Grid: Defense Use Case*. Washington, DC: E-ISAC.

Edison Electric Institute. 2000. "Master Contract." April 25. https://www.eei.org/resources-and-media/master-contract.

Edison Electric Institute. 2021. "Leading the Way: U.S. Electric Company Investment and Innovation in Energy Storage." https://www.eei.org/-/media/Project/EEI/Documents/Issues-and-Policy/Energy-Storage/Energy_Storage_Case_Studies-062021.pdf.

Edison Electric Institute. 2022. "Letter from Shelby A. Linton-Keddie, Managing Director of State Regulatory Affairs, Edison Electric Institute, to National Academies of Science Ad Hoc Committee on Net Metering."

EERE (Office of Energy Efficiency and Renewable Energy). 2022a. "Homeowner's Guide to the Federal Tax Credit for Solar Photovoltaics." https://www.energy.gov/eere/solar/homeowners-guide-federal-tax-credit-solar-photovoltaics.

EERE. 2022b. "Soft Costs." https://www.energy.gov/eere/solar/soft-costs.

EERE. 2022c. "What Is Pumped Storage Hydropower?" https://www.energy.gov/eere/water/pumped-storage-hydropower.

EIA (U.S. Energy Information Association). 2019a. "861 Electric Data." https://www.eia.gov/electricity/data/eia861.

EIA. 2019b. "Electric Sales, Revenue, and Average Price." October. https://www.eia.gov/electricity/sales_revenue_price.

EIA. 2021a. *Battery Storage in the United States: An Update on Market Trends*, pp. 5–12. Washington, DC: Department of Energy. https://www.eia.gov/analysis/studies/electricity/batterystorage/pdf/battery_storage_2021.pdf.

EIA. 2021b. "Solar Generation Was 3% of U.S. Electricity in 2020, But We Project It Will Be 20% by 2050." https://www.eia.gov/todayinenergy/detail.php?id=50357.

EIA. 2022a. "Annual Electric Power Industry Report, Form EIA-861 Detailed Data Files." November 18. https://www.eia.gov/electricity/data/eia861.

EIA. 2022b. "Annual Energy Outlook." March 3. https://www.eia.gov/outlooks/aeo/index.php.

EIA. 2022c. "Electric Power Annual." November 7. https://www.eia.gov/electricity/annual.

EIA. 2022d. "Electricity: Form EIA-861M (formerly EIA-826) Detailed Data." https://www.eia.gov/electricity/data/eia861m.

EIA. 2022e. "Form EIA-860 Detailed Data with Previous Form Data (EIA-860A/860B)." June 2. https://www.eia.gov/electricity/data/eia860.

EIA. 2022f. "Short-term Energy Outlook." https://www.eia.gov/outlooks/steo/report/natgas.php. Quantifying the Multiple Benefits of Energy Efficiency and Renewable Energy: A Guide for State and Local Governments

EPA (U.S. Environmental Protection Agency). 2018. "Part Two: Chapter Five: Estimating the Economic Benefits of Energy Efficiency and Renewable Energy." in *Quantifying the Multiple Benefits of Energy Efficiency and Renewable Energy—A Guide for State and Local Governments*. https://www.epa.gov/statelocalenergy/quantifying-multiple-benefits-energy-efficiency-and-renewable-energy-guide-state.

EPA. 2022. "Sector-Based PM2.5 Benefit per Ton Estimates." https://www.epa.gov/benmap/sector-based-pm25-benefit-ton-estimates.

EPA. 2023a. "CO-Benefits Risk Assessment Health Impacts Screening and Mapping Tool (COBRA)." https://www.epa.gov/cobra.

EPA. 2023b. "Greenhouse Gas Reduction Fund," https://www.epa.gov/inflation-reduction-act/greenhouse-gas-reduction-fund.

Erdener, B.C., C. Feng, K. Doubleday, A. Florita, and B.M. Hodge. 2022. "A Review of Behind-the-Meter Solar Forecasting." *Renewable and Sustainable Energy Reviews* 160:112224.

Fann, N., A.D. Lamson, S.C. Anenberg, K. Wesson, D. Risley, and B.J. Hubbell. 2012. "Estimating the National Public Health Burden Associated with Exposure to Ambient PM2.5 and Ozone." *Risk Analysis: An International Journal* 32(1):81–95.

Faruqui, A., R. Hledik, and S. Sergici. 2019. "A Survey of Residential Time-of-Use (TOU) Rates." The Brattle Group, Inc. https://www.brattle.com/wp-content/uploads/2021/05/17904_a_survey_of_residential_time-of-use_tou_rates.pdf.

Faruqui, A., N. Powers, and S. Sergici. 2020. *Study by Brattle Economists Evaluates Time-of-Use (TOU) Pilots for Maryland Utilities*. Prepared for Maryland Joint Utilities. Brattle Group. https://www.brattle.com/insights-events/publications/study-by-brattle-economists-evaluates-time-of-use-tou-pilots-for-maryland-utilities.

FERC (Federal Energy Regulatory Commission). 2016. *Federal Energy Regulatory Commission v. Electric Power Supply Association, et al.* U.S. Supreme Court, No. 14-840. https://www.justice.gov/sites/default/files/osg/briefs/2015/01/28/ferc_v._epsa_app.pdf.

FERC. 2020a. "2020 Assessment of Demand Response and Advanced Metering." Staff report. https://cms.ferc.gov/sites/default/files/2020-12/2020%20Assessment%20of%20Demand%20Response%20and%20Advanced%20Metering_December%202020.

FERC. 2020b. "FERC Opens Wholesale Markets to Distributed Resources: Landmark Action Breaks Down Barriers to Emerging Technologies, Boosts Competition." September 17. https://www.ferc.gov/news-events/news/ferc-opens-wholesale-markets-distributed-resources-landmark-action-breaks-down.

FERC. 2020c. *Order Dismissing Petition for Declaratory Order*. 172 FERC 61,042. New England Ratepayers Association, Docket No. EL20-42-000. https://www.ferc.gov/sites/default/files/2020-08/EL20-42-Order.pdf.

FERC. 2021. "Electric Power Markets." https://www.ferc.gov/electric-power-markets.

FERC. 2022. "FERC Acts on First of Order No. 2222 Compliance Filings." June 16. https://ferc.gov/news-events/news/ferc-acts-first-order-no-2222-compliance-filings.

Fisher, Sheehan & Colton. 2013. "The 2012 Home Energy Affordability Gap." www.homeenergyaffordabilitygap.com.

Forrester, S., G. Barbose, and C. Miller. 2022. "Private vs. Public Value of U.S. Residential Battery Storage Operated for Solar Self-Consumption." *iScience* 25(8):104714.

Forrester, S., G. Barbose, E. O'Shaughnessy, N. Darghouth, and C.C. Montañés. 2022. *Residential Solar-Adopter Income and Demographic Trends: November 2022 Update*. Berkeley, CA: Lawrence Berkeley National Laboratory.

Fox-Penner, P. 2020. *Power After Carbon: Building a Clean, Resilient Grid*. Cambridge, MA: Harvard University Press. https://www.hup.harvard.edu/catalog.php?isbn=9780674241077.

Funkhouser, E., G. Blackburn, C. Magee, and V. Rai. 2015. "Business Model Innovations for Deploying Distributed Generation: The Emerging Landscape of Community Solar in the US." *Energy Research & Social Science* 10:90–101.

Gao, X., and S. Zhou. 2022. "Solar Adoption Inequality in the U.S.: Trend, Magnitude, and Solar Justice Policies." *Energy Policy* 169(October):113–163.

Gillard, R., C. Snell, and M. Bevan. 2017. "Advancing an Energy Justice Perspective of Fuel Poverty: Household Vulnerability and Domestic Retrofit Policy in the United Kingdom." *Energy Research & Social Science* 29:53–61.

Gilliam, R. 2013. "Net Metering," p. 5, PowerPoint presentation. *The Cost, Benefits and Equity Issues of Net Energy Metering*. February 5. Washington, DC: NARUC Winter Committee Meetings.

Gillingham, K., and J.L. Sweeney. 2010. "Chapter 5: Market Failures and the Structure of Externalities," in *Harnessing Renewable Energy in Electric Power Systems: Theory, Practice, Policy*. New York: Earthscan.

Glemarec, Y., W. Rickerson, and O. Waissbein. 2012. *Transforming On-Grid Renewable Energy Markets. A Review of UNDP-GEF Support for Feed-In Tariffs and Related Price and Market-Access Instruments*. Washington, DC: Office of Scientific and Technical Information, Department of Energy. https://www.osti.gov/biblio/22090461.

Glick, D., M. Lehrman, and O. Smith. 2014. *Rate Design for the Distribution Edge: Electricity Pricing for a Distributed Resource Future*. Boulder, CO: Electricity Innovation Lab, Rocky Mountain Institute. https://rmi.org/wp-content/uploads/2017/04/2014-25_eLab-RateDesignfortheDistributionEdge-Full-highres-1.pdf.

GPSC (Georgia Public Service Commission). 2017. "A Framework for Determining the Costs and Benefits of Renewable Resources in Georgia." GPSC Docket No. 40161, Document Filing #167588. https://psc.ga.gov/search/facts-document/?documentId=167588.

GPSC. 2020. "Order Approving the Renewable and Non-Renewable Resources Tariff (RNR-10) with Modifications." GPSC Docket No. 42516, Document Filing #181467. https://psc.ga.gov/search/facts-document/?documentId=181467.

Graff, M., S. Carley, D.M. Konisky, and T. Memmott. 2021. "Which Households Are Energy Insecure? An Empirical Analysis of Race, Housing Conditions, and Energy Burdens in the United States." *Energy Research & Social Science* 79:102144.

Grafman, L., and J. Pearce. 2021. *To Catch the Sun*. Arcata, CA: Humboldt State University Press.

Green, R., and I. Staffell. 2017. "'Prosumage' and the British Electricity Market." *Economics of Energy & Environmental Policy* 6(1):33–50.

Griskevicius, V., J.M. Tybur, and B. Van den Bergh. 2010. "Going Green to Be Seen: Status, Reputation, and Conspicuous Conservation." *Journal of Personality and Social Psychology* 98:392–404.

Gundlach, J. 2020. "Climate Risks Are Becoming Legal Liabilities for the Energy Sector." *Nature Energy* 5:94–97.

Hammerstrom, D.J., R. Ambrosio, T.A. Carlon, J.G. DeSteese, G.R. Horst, R. Kajfasz, L.L. Kiesling, et al. 2007. *Pacific Northwest GridWise Testbed Demonstration Projects: Part I. Olympic Peninsula Project*. Richland, WA: Pacific Northwest National Laboratory. https://www.pnnl.gov/main/publications/external/technical_reports/PNNL-17167.pdf.

Harlan, S.L., J.H. Declet-Barreto, W.L. Stefanov, and D.B. Petitti. 2013. "Neighborhood Effects on Heat Deaths: Social and Environmental Predictors of Vulnerability in Maricopa County, Arizona." *Environmental Health Perspective* 121(2):197–204.

Hayibo, K.S., and J.M. Pearce. 2021. "A Review of the Value of Solar Methodology with a Case Study of the U.S. VOS." *Renewable and Sustainable Energy Reviews* 137:110599.

Heeter, J., A. Sekar, E. Fekete, M. Shah, and J.J. Cook. 2021a. *Affordable and Accessible Solar for All: Barriers, Solutions, and On-Site Adoption Potential*. Golden, CO: National Renewable Energy Laboratory. https://www.nrel.gov/docs/fy21osti/80532.pdf.

Heeter, J., K. Xu, and G. Chan. 2021b. *Sharing the Sun: Community Solar Deployment, Subscription Savings, and Energy Burden Reduction*. NREL/PR-6A20-80246. Golden, CO: National Renewable Energy Laboratory. https://www.nrel.gov/docs/fy21osti/80246.pdf.

Heidari, N., and J.M. Pearce. 2016. "A Review of Greenhouse Gas Emission Liabilities as the Value of Renewable Energy for Mitigating Lawsuits for Climate Change Related Damages." *Renewable & Sustainable Energy Reviews* 55:899–908.

Hempling, S., C. Elefant, K. Cory, and K. Porter. 2010. *Renewable Energy Prices in State-Level Feed-In Tariffs: Federal Law Constraints and Possible Solutions*. Technical Report NREL/TP-6A2-47408. Golden, CO: National Renewable Energy Laboratory. https://www.nrel.gov/docs/fy10osti/47408.pdf.

Hernández, D. 2016. "Understanding 'Energy Insecurity' and Why It Matters to Health." *Social Science & Medicine* 167:1–10.

Hernández, D., Y. Aratani, and Y. Jiang. 2014. *Energy Insecurity Among Families with Children*. New York: National Center for Children in Poverty, Columbia University.

Hoen, B., J. Rand, and S. Elmallah. 2019. *Commercial PV Property Characterization: An Analysis of Solar Deployment Trends in Commercial Real Estate*. Berkeley, CA: Lawrence Berkeley National Laboratory. https://emp.lbl.gov/publications/commercial-pv-property.

Holland, S.P., M.J. Kotchen, E.T. Mansur, and A.J. Yates. 2022. "Why Marginal CO_2 Emissions Are Not Decreasing for US Electricity: Estimates and Implications for Climate Policy." *Proceedings of the National Academy of Sciences* 119(8):e2116632119.

Horowitz, S.H., A.G. Phadke, and C.F. Henville. 2022. *Power System Relaying*. 5th ed. Hoboken, NJ: John Wiley and Sons.

Howell, S., Y. Rezgui, J.L. Hippolyte, B. Jayan, and H. Li. 2017. "Towards the Next Generation of Smart Grids: Semantic and Holonic Multi-Agent Management of Distributed Energy Resources." *Renewable and Sustainable Energy Reviews* 77:193–214.

ICF Consulting. 2018. *Review of Recent Cost-Benefit Studies Related to Net Metering and Distributed Solar*. Washington, DC: ICF Consulting. https://www.energy.gov/sites/default/files/2020/06/f75/ICF%20NEM%20Meta%20Analysis_Formatted%20FINAL_Revised%208-27-18.pdf.

IEA (International Energy Agency). 2017. *Digitalisation and Energy*. Paris: IEA. https://www.iea.org/reports/digitalisation-and-energy.

IEA. 2022. "Trends in Photovoltaic Applications 2022." Photovoltaic Power Systems Program. Report No. IEA PVPS T1-43:2022. https://iea-pvps.org/trends_reports/trends-2022.

Impram, S., S.V. Nese, and B. Oral. 2020. "Challenges of Renewable Energy Penetration on Power System Flexibility: A Survey." *Energy Strategy Reviews* 31:100539.

Interagency Working Group on Social Cost of Greenhouse Gases, United States Government. 2016. "Technical Support Document: Technical Update of the Social Cost of Carbon for Regulatory Impact Analysis—Under Executive Order 12866." https://www.epa.gov/sites/default/files/2016-12/documents/sc_co2_tsd_august_2016.pdf.

Institute of Policy Integrity. 2017. *The Social Cost of Greenhouse Gases and State Policy*. New York: Institute for Policy Integrity.

IREC (Interstate Renewable Energy Council). 2019. "Solar Jobs Census 2019." https://irecusa.org/resources/national-solar-jobs-census-2019.

IREC. 2022. "National Solar Jobs Census 2021." https://irecusa.org/resources/national-solar-jobs-census-2021.

Jain, H., G. Seo, E. Lockhart, V. Gevorgian, and B. Kroposki. 2020. *Blackstart of Power Grids with Inverter-Based Resources*. Preprint. NREL/CP-5D00-75327. Golden, CO: National Renewable Energy Laboratory. https://www.nrel.gov/docs/fy20osti/75327.pdf.

Johnson, E., R. Beppler, C. Blackburn, B. Staver, M. Brown, and D. Matisoff. 2017. "Peak Shifting and Cross-Class Subsidization: The Impacts of Solar PV on Changes in Electricity Costs." *Energy Policy* 106:436–444.

Joskow, P.L. 2020. "Transmission Capacity Expansion Is Needed to Decarbonize the Electricity Sector Efficiently." *Joule* 4(1):1–3.

Junlakarn, S., R. Diewvilai, and K. Audomvongseree. 2022. "Stochastic Modeling of Renewable Energy Sources for Capacity Credit Evaluation." *Energies* 15(14):5103.

Kaffine, D.T., J.M. Brannin, and J. Lieskovsky. "Emissions Savings from Wind Power Generation in Texas." *The Energy Journal* 34(1):155–175.

Kangping L., F. Wang, Z. Mi, M. Fotuhi-Firuzabad, N. Duić, and T. Wang. 2019. "Capacity and Output Power Estimation Approach of Individual Behind-the-Meter Distributed Photovoltaic System for Demand Response Baseline Estimation." *Applied Energy* 253:113595.

Kantamneni, A., R. Winkler, L. Gauchia, and J.M. Pearce. 2016. "Emerging Economic Viability of Grid Defection in a Northern Climate Using Solar Hybrid Systems." *Energy Policy* 95:378–389.

Katalenich S.M., and M.Z. Jacobson. 2021. "Renewable Energy and Energy Storage to Offset Diesel Generators at Expeditionary Contingency Bases." *The Journal of Defense Modeling and Simulation* online, https://doi.org/10.1177/15485129211051377.

Katzenstein, W., and J. Apt. 2009. "Air Emissions Due to Wind and Solar Power." *Environmental Science & Technology* 43(2):253–258.

Khodayar, M., and M. Shahidehpour. 2013. "Cutting Campus Energy Costs with Hierarchical Control: The Economical and Reliable Operation of a Microgrid." *IEEE Electrification Magazine* 1(1):40–56.

Kiesling, L. 2022. *Digitalization in Electric Power System and Regulation: A Primer*. Knowledge Problem LLC. https://pubs.naruc.org/pub/17AB6931-1866-DAAC-99FB-8D696CC1944A.

King, Jr., M.L. 1963. "Letter from Birmingham Jail." April 1.

Klee, R.J., and S. Baldinger. 2021. "Review of State Public Utility Commission Statutory Mandates." Report for the Institute for Market Transformation: Report for the Institute for Market Transformation PUC Mandate Project." New Haven, CT: Center for Business and the Environment at Yale. https://drive.google.com/file/d/1bvQTifqgxGS5i7_CU1uj0HLPBQW5kjYV/view.

Knowlton, K., M. Rotkin-Ellman, G. King, H.G. Margolis, D. Smith, G. Solomon, R. Trent, and P. English. 2009. "The 2006 California Heat Wave: Impacts on Hospitalizations and Emergency Department Visits." *Environmental Health Perspective* 117(1):61–67.

Kontokosta, C.E., V.J. Reina, and B. Bonczak. 2020. "Energy Cost Burdens for Low-Income and Minority Households." *Journal of the American Planning Association* 86(1):89–105.

Krane, J. 2017. "Climate Change and Fossil Fuel: An Examination of Risks for the Energy Industry and Producer States." *MRS Energy & Sustainability* 4:E2.

Kubli, M. 2018. "Squaring the Sunny Circle? On Balancing Distributive Justice of Power Grid Costs and Incentives for Solar Prosumers." *Energy Policy* 114:173–188.

Lane, C. 2023. "Massachusetts SMART Program Explained." *Solar Reviews*, January 11. https://www.solarreviews.com/blog/massachusetts-smart-program-replaces-srecs.

Lang, M., R. Lane, K. Zhao, S. Tham, K. Woolfe, and R. Raven. 2021. "Systematic Review: Landlords' Willingness to Retrofit Energy Efficiency Improvements." *Journal of Cleaner Production* 303:127041.

Laws, N.D., B.P. Epps, S.O. Peterson, M.S. Laser, and G.K. Wanjiru. 2017. "On the Utility Death Spiral and the Impact of Utility Rate Structures on the Adoption of Residential Solar Photovoltaics and Energy Storage." *Applied Energy* 185:627–641.

Lawson, A.J. 2019. *Net Metering: In Brief.* CRS Report R46010. Washington, DC: Congressional Research Service.

Lazar, J. 2016. *Electricity Regulation In the US: A Guide.* 2nd ed. Montpelier, VT: The Regulatory Assistance Project. https://www.raponline.org/wp-content/uploads/2016/07/rap-lazar-electricity-regulation-US-june-2016.pdf.

Leon, W., C. Farley, N. Hausman, B. Herbert, N.H. Hammer, B. Paulos, T. Reames, R. Sanders, L. Schieb, D. Deane-Ryan, and R. Navarra. 2019. *Solar with Justice: Strategies for Powering Up Under-Resourced Communities and Growing an Inclusive Solar Market.* Montpelier, VT: Clean Energy States Alliance.

Lisska et al. 2021. "Equitable Transit Oriented Development." Transportation Research Board Committee Meeting. June 8, 2022. https://www.nationalacademies.org/event/06-08-2022/data-metrics-and-analytic-methods-for-assessing-equity-impacts-of-surface-transportation-funding-programs-meeting-6.

Littell, D., C. Kadoch, P. Baker, R. Bharvirkar, M. Dupuy, B. Hausauer, C. Linvill, J. Migden-Ostrander, J. Rosenow, W. Zuan, O. Zinaman, and J. Logan. 2017. *Next-Generation Performance-Based Regulation: Emphasizing Utility Performance to Unleash Power Sector Innovation.* NREL/TP-6A50-68512. Golden, CO: National Renewable Energy Laboratory.

Louie, E.P., and J.M. Pearce. 2016. "Retraining Investment for U.S. Transition from Coal to Solar Photovoltaic Employment." *Energy Economics* 57:295–302.

Lukanov, B.R., and E.M. Krieger. 2019. "Distributed Solar and Environmental Justice: Exploring the Demographic and Socio-Economic Trends of Residential PV Adoption in California." *Energy Policy* 134:110935.

Lukanov, B., A. Makhijani, K. Shetty, Y. Kinkhabwala, A. Smith, and E. Krieger. 2022. *Pathways to Energy Affordability in Colorado.* Oakland, CA: PSE Healthy Energy, and Takoma Park, MD: Institute for Energy and Environmental Research. https://www.psehealthyenergy.org/wp-content/uploads/2022/01/Colorado-Energy-Affordability-Study_Full-Report.pdf.

Mai, T.T., P.H. Nguyen, Q.-T. Tran, A. Cagnano, G. De Carne, Y. Amirat, A-T. Le, and E. De Tuglie. 2021. "An Overview of Grid-Edge Control with the Digital Transformation." *Electrical Engineering* 103:1989–2007.

Mantegna, G., A. Burdick, S. Price, A. Olson, M.M. Kinnon, and S. Samuelsen. 2021. *Quantifying the Air Quality Impacts of Decarbonization and Distributed Energy Programs in California.* San Francisco, CA: E3, and Irvine, CA: Advanced Power and Energy Program. https://www.ethree.com/wp-content/uploads/2022/01/CPUC-Air-Quality-Report-FINAL.pdf.

Markiewicz, D., B. Wechtenhiser, J. Brendlinger, M. Campbell, M. Tims, and J. Yanuck. 2018. "Vehicle-to-Grid Testing and Demonstration Final Report." CEC-500-2018-024. Prepared for California Energy Commission by Concurrent Technologies Corporation. https://www.energy.ca.gov/sites/default/files/2021-06/CEC-500-2018-024.pdf.

Martin, M. 2021. *For the First Time, Census Bureau Data Show Impact of Geography, Income on Broadband Internet Access.* Washington, DC: U.S. Census Bureau. https://www.census.gov/library/stories/2018/12/rural-and-lower-income-counties-lag-nation-internet-subscription.html.

Massachusetts Institute of Technology Energy Initiative. 2022. *The Future of Energy Storage.* Cambridge, MA: Massachusetts Institute of Technology. https://energy.mit.edu/publication/the-future-of-energy-storage.

Massetti, E., M. Brown, M. Lapsa, I. Sharma, J. Bradbury, C. Cunliff, and Y. Li. 2017. *Environmental Quality and the U.S. Power Sector: Air Quality, Water Quality, Land Use and Environmental Justice.* ORNL/SPR-2016/772. Oak Ridge, TN: Oak Ridge National Laboratory. http://info.ornl.gov/sites/publications/files/Pub60561.pdf.

Matevosyan, J., B. Badrzadeh, T. Prevost, E. Quitmann, D. Ramasubramanian, H. Urdal, S. Achilles, J. MacDowell, S.H. Huang, V. Vital, and J. O'Sullivan. 2019. "Grid-Forming Inverters: Are They the Key for High Renewable Penetration?" *IEEE Power and Energy Magazine* 17(6):89–98.

Mateyka, P.J., and C.R. Mazur. 2021. *Homeownership in the United States: 2005 to 2019*. Washington, DC: U.S. Census Bureau. https://www.census.gov/content/dam/Census/library/publications/2021/acs/acsbr-010.pdf.

McDermott, T.E., K. McKenna, M. Heleno, B.A. Bhatti, M. Emmanuel, and S. Forrester. 2022. *Distribution System Research Roadmap*. Report No. PNNL-31550. Richland, WA: Pacific Northwest National Laboratory.

Melvin, J. 2018. "The Split Incentives Energy Efficiency Problem: Evidence of Underinvestment by Landlords." *Energy Policy* 115:342–352.

Memmott, T., S. Carley, M. Graff, and D.M. Konisky. 2021. "Sociodemographic Disparities in Energy Insecurity Among Low-Income Households Before and During the COVID-19 Pandemic." *Nature Energy* 6:186–193.

Mendonça, M., and D. Jacobs. 2009. "Feed-In Tariffs Go Global: Policy in Practice." Renewable Energy World. https://www.renewableenergyworld.com/baseload/feed-in-tariffs-go-global-policy-in-practice.

Middlemiss, L., and R. Gillard. 2015. "Fuel Poverty from the Bottom-Up: Characterising Household Energy Vulnerability Through the Lived Experience of the Fuel Poor." *Research & Social Science* 6:146–154.

Migden-Ostrander, J. 2017. "Utility 'Shadow Billing' Can Shed Light on Rate Options." *RAP*, August 29. https://www.raponline.org/blog/utility-shadow-billing-can-shed-light-on-rate-options.

Mills, A., R. Wiser, M. Milligan, and M. O'Malley. 2009. "Comment on 'Air Emissions Due to Wind and Solar Power.'" *Environmental Science & Technology* 43(15):6106–6107.

Moezzi, M., A. Ingle, L. Lutzenhiser, and B. Sigrin. 2017. *A Non-Modeling Exploration of Residential Solar Photovoltaic (PV) Adoption and Non-Adoption*. Golden, CO: National Renewable Energy Laboratory. https://www.nrel.gov/docs/fy17osti/67727.pdf.

Mongird, K., V. Viswanathan, J. Alam, C. Vartanian, V. Sprenkle, and R. Baxter. 2020. *2020 Grid Energy Storage Technology Cost and Performance Assessment*. Publication No. DOE/PA-0204. Richland, WA: Pacific Northwest National Laboratory, and Somerville, MA: Mustang Prairie Energy. https://www.pnnl.gov/sites/default/files/media/file/Final%20-%20ESGC%20Cost%20Performance%20Report%2012-11-2020.pdf.

Mundada, A.S., Y. Nilsiam, and J.M. Pearce. 2016. "A Review of Technical Requirements for Plug and Play Solar Photovoltaic Microinverter Systems in the United States." *Solar Energy* 135:455–470.

Mundada, A.S., E.W. Prehoda, and J.M. Pearce. 2017. "U.S. Market for Solar Photovoltaic Plug-and-Play Systems." *Renewable Energy* 103:255–264.

Murray, B. 2019. "The Paradox of Declining Renewable Costs and Rising Electricity Prices." *Forbes*, June 17, https://www.forbes.com/sites/brianmurray1/2019/06/17/the-paradox-of-declining-renewable-costs-and-rising-electricity-prices.

NARUC (National Association of Regulatory Utility Commissioners). 2016. *Distributed Energy Resources Rate Design and Compensation*, p. 67. Washington, DC: NARUC. https://pubs.naruc.org/pub.cfm?id=19FDF48B-AA57-5160-DBA1-BE2E9C2F7EA0.

NARUC. 2022. "Task Force Cohort Roadmaps." https://www.naruc.org/taskforce/resources-for-action/roadmaps.

NASEM (National Academies of Sciences, Engineering, and Medicine). 2017. *Enhancing the Resilience of the Nation's Electricity System*. Washington, DC: The National Academies Press. https://doi.org/10.17226/24836.

NASEM. 2017. *Valuing Climate Damages: Updating Estimation of the Social Cost of Carbon Dioxide*. Washington, DC: The National Academies Press. https://doi.org/10.17226/24651.

NASEM. 2021a. *Accelerating Decarbonization of the U.S. Energy System*. Washington, DC: The National Academies Press. https://doi.org/10.17226/25932.

NASEM. 2021b. *The Future of Electric Power in the United States*. Washington, DC: The National Academies Press. https://doi.org/10.17226/25968.

National Governors Association. n.d. "State Energy Efficiency Policy in a New Era: A Toolkit for Governors." https://www.nga.org/wp-content/uploads/2021/10/Energy-Efficiency-Toolkit.pdf.

NC-CETC (North Carolina Clean Energy Technology Center). 2021. "The 50 States of Solar." https://nccleantech.ncsu.edu/the-50-states-reports.

NC-CETC. 2022a. "Database of State Incentives for Renewables & Efficiency® (DSIRE)." https://www.dsireusa.org.

NC-CETC. 2022b. "The 50 States of Solar: Q4 2021 Quarterly Report and 2021 Annual Review." https://nccleantech.ncsu.edu/wp-content/uploads/2022/01/Q4-21-Solar-Exec-Summary-Final.pdf.

NC-CETC. 2022c. "The 50 States of Solar: Q3 2022 Report." https://www.dsireinsight.com/publications.

NC-CETC. 2023. "The 50 States of Solar: 2022 Policy Review and Q4 2022 Quarterly Report." https://nccleantech.ncsu.edu/the-50-states-reports.

NCSL (National Conference of State Legislatures). 2021. "State Renewable Portfolio Standards and Goals." https://www.ncsl.org/energy/state-renewable-portfolio-standards-and-goals.

NERC (North American Electric Reliability Corporation). 2011. *Balancing and Frequency Control: A Technical Document Prepared by the NERC Resources Subcommittee.* Princeton, NJ: NERC. https://www.nerc.com/comm/OC/BAL0031_Supporting_Documents_2017_DL/NERC%20Balancing%20and%20Frequency%20Control%20040520111.pdf.

NESP (National Energy Screening Project). 2022a. "Energy Equity and BCA." https://www.nationalenergyscreeningproject.org/resources/energy-equity-and-bca.

NESP. 2022b. "The National Standard Practice Manual (NSPM)." https://www.nationalenergyscreeningproject.org/national-standard-practice-manual.

NETL (National Energy Technology Laboratory). 2021. "Energy Technology Systems and Market Analyses." https://netl.doe.gov/sites/default/files/netl-file/21TPG_Brewer.pdf.

NIST (National Institute of Standards and Technology). 2010. *Guidelines for Smart Grid Cyber Security*, pp. 1–3. NISTIR-7628. Washington, DC: NIST. https://nvlpubs.nist.gov/nistpubs/ir/2010/NIST.IR.7628.pdf.

NIST. 2022. *Engineering Statistics Handbook: 6.4. Introduction to Time Series Analysis.* Washington, DC: NIST. https://www.itl.nist.gov/div898/handbook/pmc/section4/pmc4.htm.

Nord, M., and L.S. Kantor. 2006. "Seasonal Variation in Food Insecurity Is Associated with Heating and Cooling Costs Among Low-Income Elderly Americans." *Journal of Nutrition* 136:2939–2944.

Norström, A.V., C. Cvitanovic, and M.F. Löf. 2020. "Principles for Knowledge Co-Production in Sustainability Research." *Nature Sustainability* 3:182–190.

Norton, B.G. 2005. *Sustainability: A Philosophy of Adaptive Ecosystem Management.* Chicago, IL: University of Chicago Press.

NRC (National Research Council). 2010. *Global Sources of Local Pollution: An Assessment of Long-Range Transport of Key Air Pollutants to and from the United States.* Washington, DC: The National Academies Press. https://doi.org/10.17226/12743.

NREL (National Renewable Energy Laboratory). 2018. "Valuing the Resilience Provided by Solar and Battery Energy Storage Systems." https://www.energy.gov/sites/prod/files/2018/03/f49/Valuing-Resilience.pdf.

NREL. 2020. "PVWatts® Calculator." https://pvwatts.nrel.gov.

NREL. 2022a. "Annual Technology Baseline (ATB): 2022 Electricity ATB Technologies." https://atb.nrel.gov/electricity/2022/technologies.

NREL. 2022b. "Energy Analysis, Voluntary Green Power Procurement." https://www.nrel.gov/analysis/green-power.html.

NREL. n.d. "An Introduction to Retail Electricity Choice in the United States." https://www.nrel.gov/docs/fy18osti/68993.pdf.

NRRI (National Regulatory Research Institute). 2022a. "PURPA Tracker." https://www.naruc.org/nrri/nrri-activities/purpa-tracker.

NRRI. 2022b. "State Clean Energy Policy Tracker." https://www.naruc.org/nrri/nrri-activities/clean-energy-tracker.

NYSERDA (New York State Energy Research and Development Authority). 2022a. "New York State Releases Draft Disadvantaged Communities Criteria to Advance Climate Justice." March 9. https://www.nyserda.ny.gov/About/Newsroom/2022-Announcements/2022-03-09-New-York-State-Releases-Draft-Disadvantaged-Communities-Criteria.

NYSERDA. 2022b. "Solar Value Stack Calculator." https://www.nyserda.ny.gov/All-Programs/NY-Sun/Contractors/Value-of-Distributed-Energy-Resources/Solar-Value-Stack-Calculator.

NYSERDA. 2022c. "The Value Stack." https://www.nyserda.ny.gov/All-Programs/ny-sun/contractors/value-of-distributed-energy-resources.

OECD (Organisation for Economic Co-operation and Development). 2022. "Renewable Energy Feed-In Tariffs." https://stats.oecd.org/Index.aspx?DataSetCode=RE_FIT.

Ormandy, D., and Ezratty, V. 2012. "Health and Thermal Comfort: From WHO Guidance to Housing Strategies." *Energy Policy* 49:116–121.

O'Shaughnessy, E., and K. Ardani. 2020. "Distributed Rate Design: A Review of Early Approaches and Practical Considerations for Value of Solar Tariffs." *The Electricity Journal* 33(3):106713.

O'Shaughnessy, E., G. Barbose, R. Wiser, S. Forrester, and N. Darghouth. 2020 "The Impact of Policies and Business Models on Income Equity in Rooftop Solar Adoption." *Nature Energy* 6(1):84–91.

Pacific Northwest National Laboratory. "Energy Equity; PNNL's Vision Statement for Equity in the Power Grid." https://www.pnnl.gov/projects/energy-equity.

Pager, D., and H. Shepherd. 2008. "The Sociology of Discrimination: Racial Discrimination in Employment, Housing, Credit, and Consumer Markets." *Annual Review of Sociology* 34:181–209.

Palmintier, B., R. Broderick, B. Mather, M. Coddington, K. Baker, F. Ding, M. Reno, M. Lave, and A. Bharatkumar. 2016. "On the Path to SunShot: Emerging Issues and Challenges in Integrating Solar with the Distribution System." NREL/TP-5D00-65331. Golden, CO: National Renewable Energy Laboratory. http://www.nrel.gov/docs/fy16osti/65331.pdf.

Patel, S., and G. Reitenbach. 2010. "The Feed-In Tariff Factor." *Power*, September 1. https://www.powermag.com/the-feed-in-tariff-factor.

Paul, I., P. Howard, and J.A. Schwartz. 2017. *The Social Cost of Greenhouse Gases and State Policy*. New York: Institute for Policy Integrity, New York University School of Law. https://policyintegrity.org/files/publications/SCC_State_Guidance.pdf.

Paulos, B., S. Forrester, E. O'Shaughnessy, C. Dyson, G.L. Barbose, and R.H. Wiser. 2021. *An Assessment of Evaluation Practices of Low- and Moderate-Income Solar Programs*. Berkeley, CA: Lawrence Berkeley National Laboratory.

Pearce, J.M., and P.J. Harris. 2007. "Reducing Greenhouse Gas Emissions by Inducing Energy Conservation and Distributed Generation from Elimination of Electric Utility Customer Charges." *Energy Policy* 35:6514–6525.

Peffley, T.B., and J.M. Pearce. 2020. "The Potential for Grid Defection of Small and Medium Sized Enterprises Using Solar Photovoltaic, Battery and Generator Hybrid Systems." *Renewable Energy* 148:193–204.

Penn, I. 2022. "A Fight Over Rooftop Solar Threatens California's Climate Goals." *The New York Times*, January 24. https://www.nytimes.com/2022/01/24/business/energy-environment/california-rooftop-solar-utilities.html.

Peskoe, A. 2020. "Ari Peskoe, Director of the Electricity Law Initiative, on Comments Filed in Opposition to Net Metering Petition, June 10, 2020." Environmental and Energy Law Program, Harvard Law School. http://eelp.law.harvard.edu/wp-content/uploads/Transcript of Ari Pesko on FERC filing 6-10-20.pdf.

Peterson, D. 2021. "The Importance of Neutral Wire in 3-Phase Systems." *Control Automation*, March 21. https://control.com/technical-articles/the-importance-of-neutral-wire-in-3-phase-systems.

PJM. 2022. "Locational Marginal Emissions Introduction." https://www.pjm.com/-/media/committees-groups/forums/emerging-tech/2022/20220614/item-3---locational-marginal-emissions-introduction.ashx.

PNNL (Pacific Northwest National Laboratory). *The Distribution System Operator with Transactive (DSO+T) Study: Executive* Summary. Richland, WA: PNNL. https://www.pnnl.gov/sites/default/files/media/file/EED_1574_BROCH_DSOT-Exec-Summ_v11.pdf.

Potter, J., E. Stuart, and P. Cappers. 2018. *Barriers and Opportunities to Broader Adoption of Integrated Demand Side Management at Electric Utilities: A Scoping Study*. Berkeley, CA: Ernest Orlando Lawrence Berkeley National Laboratory. https://emp.lbl.gov/publications/barriers-and-opportunities-broader.

Prehoda, E., J.M. Pearce, and C. Schelly. 2019. "Policies to Overcome Barriers for Renewable Energy Distributed Generation: A Case Study of Utility Structure and Regulatory Regimes in Michigan." *Energies* 12(4):674.

PSE (Partnership for Southern Equity). 2021. "6 Principles of Economic Inclusion." https://psequity.org/just-opportunity.

Quackenbush, J.D. 2019. *Cost of Capital and Capital Markets Primer for Utility Regulators*. Washington, DC: National Association of Regulatory Utility Commissioners. https://pubs.naruc.org/pub.cfm?id=CAD801A0-155D-0A36-316A-B9E8C935EE4D.

Rábago, K., and Valova, R. 2018. "Revisiting Bonbright's Principles of Public Utility Rates in a DER World." *The Electricity Journal* 31(8):9–13.

Raskin, D.B., and R.L. Roberts. 2020. "Petition for Declaratory Order of New England Ratepayers Association Concerning Unlawful Pricing of Certain Wholesale Sales." United States of America Before the Federal Energy Regulatory Commission. New England Ratepayers Association, Docket No. EL20-42-000. https://s3.documentcloud.org/documents/6843679/Nera-Proposal.pdf.

Rea, J., and O. Ugwu-Oju. 2022. "Why Purchased Power Is Such a Big Deal to Better Understand US Utilities' Climate Impact." https://rmi.org/why-purchased-power-is-such-a-big-deal-to-better-understand-us-utilities-climate-impact.

Reames, T.G. 2020. "Distributional Disparities in Residential Rooftop Solar Potential and Penetration in Four Cities in the United States." *Energy Research & Social Science* 69:101612.

Regeneration. 2022. "Electrify Everything." https://regeneration.org/nexus/electrify-everything.

Reta, M. 2023. "How States Can Help Implement the Justice40 Initiative." *NRDC Expert Blog*, January 27. https://www.nrdc.org/experts/mikyla-reta/how-states-can-help-implement-justice40-initiative.

Revesz, R.L., and B. Unel. 2016. *Managing the Future of the Electricity Grid: Distributed Generation and Net Metering*. Public Law Research Paper No. 16-09. New York: New York University School of Law. https://policyintegrity.org/files/publications/Managing_the_Future_of_the_Electricity_Grid.pdf.

Revuelta, P.S., P.L. Salvador, and J.P. Thomas. 2016. "8 - Distributed Generation," pp. 285–322 in *Active Power Line Conditioners: Design, Stimulation and Implementation for Improving Power Quality*. Amsterdam: Elsevier Inc.

Richter, L. 2014. "Social Effects in the Diffusion of Solar Photovoltaic Technology in the UK." EPRG Working Paper 1332. University of Cambridge. https://www.eprg.group.cam.ac.uk/wp-content/uploads/2013/12/1332-Final.pdf.

Riveros, J.Z., M. Kubli, and S. Ulli-Beer. 2019. "Prosumer Communities as Strategic Allies for Electric Utilities: Exploring Future Decentralization Trends in Switzerland." *Energy Research & Social Science* 57:101219.

Roberts, D. 2017. "The Key to Tackling Climate Change: Electrify Everything." *Vox*, October 27. https://www.vox.com/2016/9/19/12938086/electrify-everything.

Ross, E., M. Day, C. Ivanova, A. McLeod, and J. Lockshin. 2022. "Intersections of Disadvantaged Communities and Renewable Energy Potential: Data Set and Analysis to Inform Equitable Investment Prioritization in the United States." *Renewable Energy Focus* 41:1–14.

Rossi, J. 2016. "Federalism and the Net Metering Alternative." *The Electricity Journal* 29:13–18.

Sakka, A., M. Santamouris, I. Livada, F. Nicol, and M. Wilson. 2012. "On the Thermal Performance of Low Income Housing During Heat Waves." *Energy Build* 49:69–77.

Sanchez Rodriguez, R., D. Ürge-Vorsatz, and A.S. Barau. 2018. "Sustainable Development Goals and Climate Change Adaptation in Cities." *Nature Climate Change* 8:181–183.

Sappington, D.E.M., and D.L. Weisman. 2021. "Designing Performance-Based Regulation to Enhance Industry Performance and Consumer Welfare." *Electricity Journal* 34(2):106902.

Satchwell, A., A. Mills, and G. Barbose. 2015a. "Quantifying the Financial Impacts of Net-Metered PV on Utilities and Ratepayers." *Energy Policy* 80(May):133–144.

Satchwell, A., A. Mills, and G. Barbose. 2015b. "Regulatory and Ratemaking Approaches to Mitigate Financial Impacts of Net-Metered PV on Utilities and Ratepayers." *Energy Policy* 85:115–125.

Scavo, J., S. Korosec, E. Guerrero, B. Pennington, and P. Doughman. 2016. *Low-Income Barriers Study, Part A: Overcoming Barriers to Energy Efficiency and Renewables for Low-Income Customers and Small Business Contracting Opportunities in Disadvantaged Communities*. Publication Number: CEC-300-2016-009-CMF. Sacramento, CA: California Energy Commission.

SEPA (Smart Electric Power Alliance). 2019. "2019 Utility Energy Storage Market Snapshot." https://sepapower.org/resource/2019-utility-energy-storage-market-snapshot.

Sexton, S.E., and A.L. Sexton. 2014. "Conspicuous Conservation: The Prius Halo and Willingness to Pay for Environmental Bona Fides." *Journal of Environmental Economics and Management* 67:303–317.

Shah, K.K., A.S. Mundada, and J.M. Pearce. 2015. "Performance of US Hybrid Distributed Energy Systems: Solar Photovoltaic, Battery and Combined Heat and Power." *Energy Conversion and Management* 105:71–80.

Shakya, S.R., I. Bajracharya, R.A. Vaidya, P. Bhave, A. Sharma, M. Rupakheti, and T.R. Bajracharya. 2022. "Estimation of Air Pollutant Emissions from Captive Diesel Generators and Its Mitigation Potential Through Microgrid and Solar Energy." *Energy Reports* 8:3251–3262.

Sitthiyot, T., and K. Holasut, K. 2020. "A Simple Method for Measuring Inequality." *Palgrave Communications* 6(112):1–9.

Smith, K.M., C. Koski, and S. Siddiki. 2021. "Regulating Net Metering in the United States: A Landscape Overview of States' Net Metering Policies and Outcomes." *The Electricity Journal* 34(2):106901.

SMUD (Sacramento Municipal Utility District) and E3. 2020. *Value of Solar and Solar + Storage*. Sacramento, CA: Sacramento Municipal Utility District and E3. https://www.smud.org/-/media/Documents/Rate-Information/2022-2023-Rate-Action/VOSstudy.ashx.

Solar Electric Power Association. 2013. "Ratemaking, Solar Value and Solar Net Energy Metering—A Primer," pp. 5–6. Washington, DC: Solar Electric Power Association. https://www.energy.gov/sites/prod/files/2015/03/f20/sepa-nem-report-0713-print.pdf.

Solar Energy Industries Association. 2022. "Solar Industry Research Data." https://www.seia.org/solar-industry-research-data.

Song, G., B. Cao, and L. Chang. 2022. "Review of Grid-Forming Inverters in Support of Power System Operation." *Chinese Journal of Electrical Engineering* 8(1):1–15.

Sovacool, B.K., and M.H. Dworkin. 2015. "Energy Justice: Conceptual Insights and Practical Applications." *Applied Energy* 142:435–444.

Speelman, L., and Y. Numata. 2022. "Harnessing the Power of S-Curves: How S-Curves Work and What We Can Do to Accelerate Them." https://rmi.org/insight/harnessing-the-power-of-s-curves.

Stanton, T. 2019. *Review of State Net Energy Metering and Successor Rate Designs*. NRRI 19-01. Silver Spring, MD: National Regulatory Research Institute. https://pubs.naruc.org/pub/A107102C-92E5-776D-4114-9148841DE66B.

Stanton, T. 2020. *Solar Energy That Pays for Low-Income Customers and Communities*. NRRI Insights Paper. Silver Spring, MD: National Regulatory Research Institute. https://pubs.naruc.org/pub/46965D7D-155D-0A36-315D-58319B591EB8.

Stanton, T., and D. Phelan. 2013. "State and Utility Solar Energy Programs: Recommended Approaches for Growing Markets." NRRI 13-07. Silver Spring, MD: National Regulatory Research Institute. https://pubs.naruc.org/pub/FA86BBED-D69F-C1EC-D308-0BB2CDAF5E37.

Stavins, R.N., T. Schatzki, and R. Scott. 2019. "Transitioning to Long-Run Effective and Efficient Climate Policies." *M-RCBG Faculty Working Paper Series* 2019-01. https://www.hks.harvard.edu/sites/default/files/centers/mrcbg/files/FWP2019-01.pdf.

Subramanian, S., W. Berg, E. Cooper, M. Waite, B. Jennings, A. Hoffmeister, and B. Fadie. 2022. "2022 State Energy Efficiency Scorecard." *American Council for an Energy-Efficiency Economy*. https://www.aceee.org/sites/default/files/pdfs/u2206.pdf.

Sunter, D.A., S. Castellanos, and D.M. Kammen. 2019. "Disparities in Rooftop Photovoltaics Deployment in the United States by Race and Ethnicity." *Nature Sustainability* 2(1):71–76.

Taylor, M., J. McLaren, K. Cory, T. Davidovich, J. Sterling, and M. Makhyoun. 2015. "Value of Solar: Program Design and Implementation Considerations." NREL/TP-6A20-62361. Golden, CO: National Renewable Energy Laboratory. https://www.nrdel.gov/docs/fy15osti/62361.pdf.

Tessum, C.W., D.A. Paolella, S.E. Chambliss, J.S. Apt, J.D. Hill, and J.D. Marshall. 2021. "PM2.5 Polluters Disproportionately and Systemically Affect People of Color in the United States." *Science Advances* 7(18):eabf4491.

Tierney, S.F. 2016. *The Value of "DER" to "D": The Role of Distributed Energy Resources in Supporting Local Electric Distribution System Reliability*. Denver, CO: Analysis Group, Inc. https://www.analysisgroup.com/globalassets/content/news_and_events/news/value_of_der_to-_d.pdf.

Tsao, S. 2017. "PJM, CAISO Detail Load Changes During Solar Eclipse." *S&P Global Market Intelligence*, September 13. https://www.spglobal.com/marketintelligence/en/news-insights/trending/h-gQiWXabDmfyAvghazbsw2.

UNEP (United Nations Environment Programme). 2012. "Feed-In Tariffs as a Policy Instrument for Promoting Renewable Energies and Green Economies in Developing Countries." https://unfccc.int/files/documentation/submissions_from_parties/adp/application/pdf/unep_us___ws2.pdf.

University of Michigan School for Environment and Sustainability. 2022. "Energy Equity Project Report." https://energyequityproject.com/wp-content/uploads/2022/08/220174_EEP_Report_8302022.pdf

van Benthem, A., K. Gillingham, and J. Sweeney. 2008. "Learning-by-Doing and the Optimal Solar Policy in California." *The Energy Journal* 29(3):131–151.

Van Nostrand, J.M. 2019. "Quantifying the Resilience Value of Distributed Energy Resources Quantifying the Resilience Value of Distributed Energy Resources." *Journal of Land Use* 35(1):15–34.

Vandentorren, S., P. Bretin, A. Zeghnoun, L. Mandereau-Bruno, A. Croisier, C. Cochet, J. Ribéron, I. Siberan, B. Declercq, and M. Ledrans. 2006. "August 2003 Heat Wave in France: Risk Factors for Death of Elderly People Living at Home." *European Journal of Public Health* 16:583–591.

Venugopalan, S., and V. Rai. 2015. "Topic Based Classification and Pattern Identification in Patents." *Technological Forecasting and Social Change* 94:236.

Walker, G., and R. Day. 2016. "Necessary Energy Uses and a Minimum Standard of Living in the United Kingdom: Energy Justice or Escalating Expectations?" *Energy Research & Social Science* 18:129–138.

Wan, Y.H., and H.J. Green. 1998. "Current Experience with Net Metering Programs." CP-500-24527. Golden, CO: National Renewable Energy Laboratory. Paper presented at WINDPOWER '98, Bakersfield, CA. https://www.nrel.gov/docs/legosti/old/24527.pdf.

Weisman, D. 2018. "A Report on the Theory and Practice of Performance-Based Regulation." *SSRN*. https://ssrn.com/abstract=3765691.

Weissman, G. 2017. "Rental Rooftops: A Missed Opportunity for Solar Energy." https://frontiergroup.org/blogs/blog/fg/rental-rooftops-missed-opportunity-solar-energy.

White, LV., and N.D. Sintov. 2020. "Health and Financial Impacts of Demand-Side Response Measures Differ Across Sociodemographic Groups." *Nature Energy* 5:50–60.

Wilkinson, F. 2022. "How Copake, New York, Fought and Then Accepted a Solar Farm." *Bloomberg*, March 20. https://www.bloomberg.com/opinion/articles/2022-03-20/how-copake-new-york-fought-and-then-accepted-a-solar-farm.

Wilson, B. 2020. "Urban Heat Management and the Legacy of Redlining." *Journal of the American Planning Association* 86(4):443–457.

Wolak, F.A. 2018. *The Evidence from California on the Economic Impact of Inefficient Distribution Network Pricing*. Cambridge, MA: National Bureau of Economic Research. https://www.nber.org/papers/w25087.

Wolske, K.S., K.T. Gillingham, and P.W. Schultz. 2020. "Peer Influence on Household Energy Behaviours." *Nature Energy* 5:202–212.

Wolske, K.S., P.C. Stern, and T. Dietz. 2017. "Explaining Interest in Adopting Residential Solar Photovoltaic Systems in the United States: Toward an Integration of Behavioral Theories." *Energy Research & Social Science* 25:134–151.

Wood, E. 2022. "Army Releases Five-Year Microgrid Implementation Plan." *Microgrid Knowledge*, October 10. https://www.microgridknowledge.com/editors-choice/article/11436859/army-releases-five-year-microgrid-implementation-plan.

Wood Mackenzie and Solar Energy Industries Association. 2022a. "US Solar Market Insight: 2021 Year-in-Review." https://www.woodmac.com/reports/power-markets-us-solar-market-insight-2021-year-in-review-150013339.

Wood Mackenzie and Solar Energy Industries Association. 2022b. "U.S. Solar Market Insight Report: Executive Summary—Q3 2022," p. 16. https://www.woodmac.com/industry/power-and-renewables/us-solar-market-insight.

Wu, K., Y.M. Zhang, R. Hardison, and E. Weber. 2021. *Business Models to Accelerate the Utilization of Distributed Energy Resources*. NREL/TP-6A20-79549. Golden, CO: National Renewable Energy Laboratory. https://www.nrel.gov/docs/fy22osti/79549.pdf.

Yazdani, H., M. Doostizadeh, and F. Aminifar. "Unlocking the Value of Flexibility of Behind-the-Meter Prosumers: An Overview of Mechanisms to Esteemed Trends." *The Electricity Journal* 35(5):107126.

Yu, J., Z. Wang, A. Majumdar, and R. Rajagopar. 2018. "DeepSolar: A Machine Learning Framework to Efficiently Construct a Solar Deployment Database in the United States." *Joule* 2(12):2605–2617.

Zachary, J. 2010. "Feed-In Tariffs Around the World." *Power*, September 1. https://www.powermag.com/feed-in-tariffs-around-the-world.

Zanocco, C., J. Flora, R. Rajagopal, and H. Boudet. 2021. "When the Lights Go Out: Californians' Experience with Wildfire-Related Public Safety Power Shutoffs Increases Intention to Adopt Solar and Storage." *Energy Research & Social Science* 79:102183.

Zhou, Q., M. Shahidehpour, A. Paaso, S. Bahramirad, A. Alabdulwahab, and A. Abusorrah. 2020. "Distributed Control and Communication Strategies in Networked Microgrids." *IEEE Communications Surveys and Tutorials* 22(4):2586–2633.

Zinaman, O.R., T. Bowen, and A.Y. Aznar. 2020. *An Overview of Behind-the-Meter Solar-Plus-Storage Regulatory Design: Approaches and Case Studies to Inform International Applications*. Report No. NREL/TP-7A40-75283. Golden, CO: National Renewable Energy Laboratory.

Zitelman, K., and J. McAdams. 2021. *The Role of State Utility Regulators in a Just and Reasonable Energy Transition—Examining Regulatory Approaches to the Economic Impacts of Coal Retirements*. Washington, DC: National Association of Regulatory Utility Commissioners. https://pubs.naruc.org/pub/952CF0F2-1866-DAAC-99FB-0C6352BF7CB0.

Appendixes

Committee Biographical Information

Janet Gail Besser (*Chair*) is an energy policy and regulatory strategy leader with a proven track record of developing and implementing frameworks, rules, and markets to advance competition, efficiency, and clean energy. She has deep regulatory expertise and experience in retail and wholesale market structures, as well as broad industry experience as a regulator, utility, developer, consultant, and consumer advocate. She has particular strength in bringing together diverse groups in pursuit of a common goal. She most recently was a vice president at Smart Electric Power Alliance, focusing on regulatory and utility business innovation. Previously, she was the executive vice president of the Northeast Clean Energy Council; vice president of regulatory strategy and policy at National Grid; chair and commissioner of the Massachusetts Department of Public Utilities; and an executive and expert at consulting firms Analysis Group and Lexecon. She held senior staff roles at the Massachusetts Energy Office and the New Hampshire Public Utilities Commission, was the policy director for a national IPP association and worked for a small hydro developer and low-income consumer advocate. Besser chairs the Power Options Board and sits on the VEIC Board where she chairs the Impact and Evaluation Committee. Besser received the 2019 Northeast Energy and Commerce Association Vanguard Award and the 2014 New England Women in Energy and Environment Achievement award. She has an MPP from the Kennedy School of Government, Harvard University, and a BA, magna cum laude, from Williams College, where she was elected to Phi Beta Kappa.

Anuradha M. Annaswamy is the founder and director of the Active-Adaptive Control Laboratory in the Department of Mechanical Engineering at the Massachusetts Institute of Technology (MIT). Her research interests span adaptive control theory and its applications to aerospace, automotive, propulsion, and energy systems as well as cyber-physical systems such as Smart Grids, Smart Cities, and Smart Infrastructures. She has received best paper awards (Axelby; CSM), as well as distinguished member and distinguished lecturer awards from the *Institute of Electrical and Electronics Engineers* (IEEE) Control Systems Society (CSS) and a Presidential Young Investigator award from the National Science Foundation. She is a fellow of IEEE and the International Federation of Automatic Control. She is the recipient of the Distinguished Alumni Award from the Indian Institute of Science for 2021. Annaswamy received her PhD in electrical engineering from Yale

University in 1985. She has served as a faculty member at Yale, Boston University, and MIT. She is the author of a graduate textbook on adaptive control, co-editor of two vision documents on smart grids as well as two editions of the *Impact of Control Technology* report, author of several journals and conference publications, and a member of the National Academies of Sciences, Engineering, and Medicine's committee that published the report *The Future of Electric Power in the United States* in 2021. She served as the president of CSS in 2020. She has been serving as a faculty lead in the Electric Power Systems workstream in the MIT Future Energy Systems Center since September 2021.

Galen Barbose is a research scientist in the Electricity Markets and Policy Department at Lawrence Berkeley National Laboratory. Barbose conducts research on issues in the electricity industry related to renewable energy and distributed energy resources (DER), with a focus on distributed solar cost and technology trends, income and demographic adoption patterns of residential solar adopters, retail rate design, utility business models, and renewable portfolio standards. As part of his research portfolio, Barbose manages three annual reports: *Tracking the Sun*, *Residential Solar Adopter Demographic Trends*, and *U.S. Renewable Portfolio Standards Annual Status Report*. Barbose also provides direct analytical support to key decision makers and stakeholders within the electricity industry. Since joining Berkeley Laboratory in 2001, Barbose has authored more than 100 publications and has presented his work on hundreds of occasions. Prior to Berkeley Laboratory, Barbose was a project engineer at Nexant and Schiller Associates, providing project management, engineering, and consulting services for utility-funded energy efficiency and other demand-side management programs. Barbose holds bachelor's degrees in mechanical engineering and philosophy from the University of California (UC), Berkeley, and UC Santa Cruz, respectively, and a master's degree from the Energy and Resources Group (ERG) at UC Berkeley.

Marilyn A. Brown (NAE/NAS) is a Regents' Professor in the School of Public Policy at the Georgia Institute of Technology and a joint faculty member at Oak Ridge National Laboratory. At Georgia Tech, she created and co-leads the Climate and Energy Policy Laboratory and the Master of Sustainable Energy and Environmental Management. Her research focuses on the design and modeling of energy and carbon reduction policies and programs, highlighting opportunities on the customer side of the electric meter. Using data analytics and energy-engineering-behavioral models, she examines technology and market transitions. Prior to Georgia Tech, she worked at Oak Ridge National Laboratory, where she deployed her expertise in scenarios focused on a clean energy future, bringing a fact-based and authoritative perspective to energy sustainability discussions, influencing policy initiatives in the United States and around the globe. Focusing on the electric utility industry, Brown has joined fundamental behavioral and social science principles with economic-engineering models. She has demonstrated why energy efficiency and affordability are as essential to energy security as the more traditional dimensions of fuel diversity and import independence. Her research on determinants of energy burden is highly cited. In 2007, Brown contributed to the Intergovernmental Panel on Climate Change reports that shared the 2007 Nobel Peace Prize. From 2010 to 2017, she served two terms as a Presidential appointee to the Board of Directors of the Tennessee Valley Authority, the nation's largest public power provider. From 2014 to 2018 she served two terms on the Department of Energy's Electricity Advisory Committee where she chaired the Smart Grid Subcommittee. She has written nine books on the clean energy transition, which have been used by researchers and students around the world. She is a member of the National Academy of Engineering and the National Academy of Sciences. She has served on eight National Academies' committees.

Mohit Chhabra is a senior scientist at the Natural Resources Defense Council (NRDC); he helps lead NRDC's research and advocacy efforts to decarbonize California's energy sector cost-effectively. Prior to joining NRDC, Chhabra worked as a contract analyst to the Northwest Power and Conservation Council conducting energy conservation assessments. Chhabra has expertise in evaluating and developing policy for DER. He developed policy proposals to develop a new metric, and a new framework to set goals and develop budgets for all distributed resources; that metric was adopted by the California Public Utilities Commission (CPUC) for energy efficiency and demand response programs. He is currently working to update California's Net Energy Metering policy at the CPUC. Chhabra is a voting member of the Regional Technical Forum (RTF) of the Pacific Northwest and is also a part of the Conservation Resources Advisory Committee of the Pacific Northwest. In 2016, the RTF recognized his contributions to energy conservation in the Pacific Northwest. He holds a master's in civil environmental and architectural engineering from the University of Colorado Boulder and a bachelor's degree in mechanical engineering from the University of Pune in India.

Elena M. Krieger is the director of research at the energy science and policy research institute Physicians, Scientists, and Engineers for Healthy Energy (PSE). She joined PSE in 2013 to launch the organization's clean energy practice area and now oversees its scientific research efforts. Her current work focuses on accelerating the transition to clean energy resources and developing transition pathways that realize health, environment, equity, and resilience co-benefits. She simultaneously works closely with community organizations, non-profits, policymakers, regulators, and other stakeholders to use science to inform energy policy. Krieger's recent and ongoing research areas include solar + storage resilience hubs; integration of health, equity, and environmental metrics into state-level deep decarbonization strategies; sociodemographic trends of solar adoption;

and identification of opportunities to replace peaker power plants with energy storage. She is a member of the Disadvantaged Communities Advisory Group to the California Energy Commission and the California Public Utilities Commission, and a member of the National Academies' New Voices in Science, Engineering, and Medicine program 2021 cohort. She received her PhD in mechanical and aerospace engineering from Princeton, where her research focused on optimizing energy storage in renewable systems, and she holds an AB in physics and astronomy and astrophysics from Harvard.

Joshua M. Pearce is the John M. Thompson Chair in Information Technology & Innovation at the Thompson Centre for Engineering Leadership & Innovation. He holds appointments at Ivey Business School, the top-ranked business school in Canada and the Department of Electrical & Computer Engineering at Western University in Canada, a top 1 percent global university. He runs the Free Appropriate Sustainability Technology research group, which concentrates on the use of open-source appropriate technology to find collaborative solutions to problems in sustainability and to reduce poverty. His research spans areas of engineering of solar photovoltaic technology, open hardware, and distributed recycling and additive manufacturing using RepRap 3D printing, but also includes policy and economics. According to Elsevier's citation metrics last year, he is in the top 0.06 percent most cited scientists globally and is continually ranked in the top 0.1 percent for his accessible research on academia.edu. He was the first Endowed Richard Witte Professor at the Michigan Technological University. He was a Fulbright-Aalto University Distinguished Chair, a visiting professor of photovoltaics and nanoengineering at Aalto University, Finland, and a visiting Professor Équipe de Recherche sur les Processus Innovatifs, Université de Lorraine, France. He received his PhD in materials engineering from The Pennsylvania State University for the development of low-cost solar cells.

Autumn F. Proudlove is the senior policy program director at the North Carolina Clean Energy Technology Center at North Carolina State University. Proudlove's professional expertise is in clean energy policy, with an emphasis on DER. Proudlove is the lead author of the Center's *50 States of Solar* quarterly report series, which tracks policy changes under consideration across the United States related to net metering, DER valuation, and utility rate design, among other topics. Proudlove received her master's degree in energy regulation and law from Vermont Law School and her bachelor's degree in environmental studies from Dartmouth College.

Varun Rai is the Walt and Elspeth Rostow Professor in the Lyndon B. Johnson School of Public Affairs at The University of Texas (UT) at Austin, with a joint appointment in mechanical engineering. Through his interdisciplinary research, delving with issues at the interface of energy systems, complex systems, decision science, and public policy, he is developing effective policy approaches to help accelerate the transition of the global energy system into a more sustainable and resilient one. He was a global economic fellow in 2009. During 2013–2015 he was a commissioner for the vertically-integrated electric utility Austin Energy, and from 2019 to 2021 he served as the director of the UT Austin Energy Institute. In 2016 the Association for Public Policy Analysis & Management awarded him the David N. Kershaw Award and Prize, which "was established to honor persons who, at under the age of 40, have made a distinguished contribution to the field of public policy analysis and management." He received his PhD and MS in mechanical engineering from Stanford University and a bachelor's degree in mechanical engineering from the Indian Institute of Technology Kharagpur.

Mohammad Shahidehpour (NAE) joined the Illinois Institute of Technology (IIT) in 1983 where he is presently a University Distinguished Professor. He also serves as the Bodine Chair Professor and the director of the Robert W. Galvin Center for Electricity Innovation at IIT. Shahidehpour has been the principal investigator of over $60 million in grants and contracts on power system operation and control, smart grid research and development, and large-scale integration of renewable energy. His Department of Energy Project on Perfect Power Systems has converted the entire IIT campus to an islandable microgrid. He has co-authored six books and over 800 papers on electric power system operation and planning and was the recipient of six best paper awards for his IEEE publications. He served as the vice president of publications for the *Institute of Electrical and Electronics Engineers* (IEEE) Power and Energy Society, editor of *IEEE Transactions on Power Systems*, and the founding editor-in-chief of the *IEEE Transactions* on Smart Grid. He is a fellow of IEEE, the American Association for the Advancement of Science, and the National Academy of Inventors (NAI); a laureate of KIA (Khwarizmi International Award); and an elected member of the National Academy of Engineering. He is listed as a highly cited researcher on the Web of Science (ranked in the top 1 percent by citations demonstrating significant influence among his peers).

Nicole D. Sintov serves as an assistant professor at The Ohio State University's School of Environment and Natural Resources. As an environmental psychologist, she strives to advance psychological theory while producing insights that can be applied to benefit the environment and human welfare. Broadly, her work focuses on the role of individual-level behavior and decision making in sustainable consumption with an emphasis on energy contexts, such as home energy use or energy-efficient technology adoption and use. Using an equity lens, her work also examines the financial, health, and social impacts of emerging energy technologies and programs (e.g., utility rate plans) on members of different vulnerable groups. In her career, she has published 25 refereed articles, 2 book chapters, 16 peer-reviewed conference proceedings, and given over 30 scholarly presentations. Her work has appeared in high-impact scholarly outlets such as *Nature Energy*, and has received media coverage by CNN, NPR, *National Geographic*, *The Guardian*, *Scientific American*, *LA Times*, among dozens of other outlets. She holds a PhD in psychology from the University of Southern California and a BS in psychology and ecology from the University of California, San Diego.

Thomas S. Stanton retired at the end of June 2022, when the National Regulatory Research Institute (NRRI) closed permanently. Stanton had been principal researcher, energy and environment, at NRRI since fall 2010. NRRI was founded by and served the National Association of Regulatory Utility Commissioners. NRRI provided regulatory policy research and educational services for agencies responsible for public utility regulation in the 50 states and U.S. territories. Stanton conducted research and writing, and provided education on public utility regulatory policy, primarily about renewable energy, energy efficiency, infrastructure modernization, industry innovation and restructuring, and climate change. Prior to joining NRRI in fall 2010, Stanton worked 10 years for the Michigan State Energy Office and then over 22 years as a member of the Michigan Public Service Commission (MPSC) staff. While on MPSC staff, Stanton served as a renewable energy section manager. He helped establish Michigan's initial voluntary net energy metering (NEM) program and then helped implement Michigan's initial legislated renewable energy portfolio standard and NEM program. Stanton provided expert witness testimony in 16 MPSC cases from 1993 to 2009. He has experience in project design and implementation, technology assessments, and program evaluations. His avocations include appropriate technology, sustainable development, and ecovillages. Stanton has been an adjunct professor at three Michigan colleges and universities, designing curricula and teaching both undergraduate and graduate-level courses. He received a BA in communications and an MA in journalism from Michigan State University and an MS in public administration from Western Michigan University.

Terrence G. Surles consults with Stanford University evaluating California's ability to meet its carbon neutrality goals. He consults with the University of California, Berkeley/ *California Institute* for Energy and Environment for the California Public Utilities Commission on expediting disputes related to the installation of behind-the-meter generation, as well as analysis and policy development for demand response, energy-efficient, and storage technologies. He consults with Hawaii Natural Energy Institute (HNEI) as part of HECO's Integrated Grid Planning process for the integration of distributed resources while maintaining grid stability and resiliency. He led the review of the Department of Energy's (DOE's) national laboratory capabilities for DOE's Grid Modernization Initiative. Previous full-time positions included the administrator for the Hawaii State Energy Office, clean energy solutions lead for the University of Hawaii College of Social Sciences, and

HNEI's technology integration and policy analysis lead. He held leadership positions at Argonne National Laboratory, Lawrence Livermore National Laboratory, and the Electric Power Research Institute. Surles has been involved and has led efforts in energy and environmental systems analysis since 1974. The more recent focus is on addressing climate change by reducing carbon emissions from electricity generation. This work includes an analysis of impacts on the grid by increasing percentages of intermittent renewable resources. He has previously served on five National Academies' committees and panels.

Susan F. Tierney works on economic, environmental, regulatory, and policy issues related to the nation's energy transitions, especially in the electric sector. She consults governments, foundations, non-governmental organizations, energy companies, energy consumers on these issues. Tierney chairs the National Renewable Energy Laboratory's External Advisory Council and serves on the boards of foundations and think tanks working on addressing climate change. She was the assistant secretary for policy at the Department of Energy (DOE), and in Massachusetts served as a cabinet officer for environmental affairs and the commissioner of the Department of Public Utilities. Her research, writing, consulting, and pro-bono activities focus on issues at the intersection of economics, environmental improvement, energy system performance, and policy design. She has written about, provided expert testimony on and participated actively in industry discussions on U.S. electric industry transitions. Tierney is a national associate of the National Research Council, is a member of the National Academies' Committee on Accelerating Decarbonization in the United States and served on the Committee on the Future of Electric Power in America. She was a visiting fellow in policy practice at the University of Chicago's Energy Policy Institute. In 2015, she received the National Association of Regulatory Utility Commissioners' Mary Kilmarx Award for Good Governance, Clean Energy and the Environment, and in 2014, received the Lifetime Achievement Award from DOE's Clean Energy Education & Empowerment (C3E) Initiative. Tierney has a PhD and master's degree in regional planning from Cornell University and a BA in art history from Scripps College.

Net Metering and Net Metering Reform Status in U.S. Jurisdictions

This appendix includes details about the status of net metering and related policies in the United States. The source for these data, unless otherwise specified, is the North Carolina Clean Energy Technology Center (NC-CETC) "50 States of Solar" series of reports.[1] Additional details and links to each jurisdiction's foundational documents are available at the NC-CETC website and on the NC-CETC Database of State Incentives for Renewables and Efficiency (DSIRE).[2] Figure B-1 shows the numbers of actions taken by state legislatures and utility regulators, from 2015 to 2022, depicting seven major categories of net metering reforms.[3]

Figure B-2 shows trends in the global price of solar panels and in global solar panel installations by year, from 1975 through 2019. The International Energy Agency (IEA) estimates that global installations nearly doubled again between 2019 and 2021 (totaling nearly 175 GW) and projected another 275 GW would be installed by year end 2022.[4] IEA shows installations over the past decade averaging about 60 percent in utility scale centralized photovoltaic (PV) power plants with 40 percent grid-connected distributed generation, and only small single-digit percentages of off-grid solar. IEA does report a major decline in solar PV installed as a result of feed-in tariffs, declining from 85 percent of all PV installations in 2010 to 60 percent in 2015 and 28 percent in 2021. Feed-in tariff auctions and purchase power agreements (PPAs) (20 percent), direct subsidies and tax breaks (16 percent), and net metering and similar incentives for self-consumption (10 percent) were also reported for almost half of global 2021 PV installations, with non-incentivized installations accounting for nearly 25 percent of 2021 system additions.[5]

---

[1] North Carolina Clean Energy Technology Center (NC-CETC). 2021 and 2022. "The 50 States of Solar Reports." Raleigh: North Carolina State University. https://nccleantech.ncsu.edu/the-50-states-reports.

[2] See https://nccleantech.ncsu.edu/our-work/energy-policy-markets/ and NC-CETA. 2022a. "Database of State Incentives for Renewables and Efficiency® (DSIRE)." https://www.dsireusa.org.

[3] Legislation is counted if it has been introduced and has passed at least one house of a state legislature. Regulatory Commission actions are counted.

[4] For additional information see https://www.iea.org/reports/solar-pv.

[5] International Energy Agency. 2022. "Trends in Photovoltaic Applications 2022." Photovoltaic Power Systems Program. Report No. IEA PVPS T1-43:2022. https://iea-pvps.org/trends_reports/trends-2022.

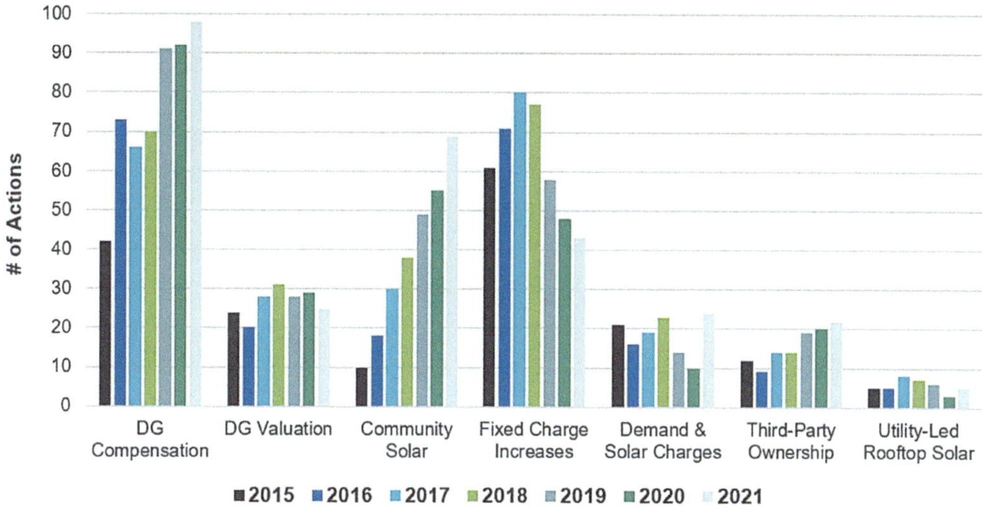

FIGURE B-1 Number of U.S. Solar Policy Actions by major policy category by year, 2015–2021.
SOURCE: North Carolina Clean Energy Technology Center, 2022b, "50 States of Solar Q4 2021 and Annual Report for 2021."

Table B-1 summarizes jurisdiction decisions about net metering reforms, from 2015 to 2022. Table B-2 summarizes decisions about net metering reforms that were made by some of the larger public power authorities in the United States, from 2012 through 2021.[6] Table B-3 provides a summary of decisions that have added specific fees for customers with distributed generation. Those decisions were made between 2017 and the time of data gathering in 2022. They are listed by state, including some that apply to only some utility companies. As noted in the "Year of Decision" column, of the 11 decisions adding charges specifically to customers with DG, or in some cases only to customers with on-site PV, five of the decisions were later overturned by state legislators, state courts, or state regulatory commissions. Table B-4 shows the different major categories of benefits and costs that have been included in over 30 value of solar (VOS) and value of distributed energy resources (VDER) studies. Table B-4 also shows that several of the benefit categories are included in nearly every study, while other benefit categories and the two cost categories were not included in many of the studies. Environmental benefits were included in all by six of the studies.

[6] NC-CETC has not compiled the details of net metering policies and changes for all municipal and publicly owned utilities in the United States but does collect and report information about the larger ones, with more than 100,000 customers.

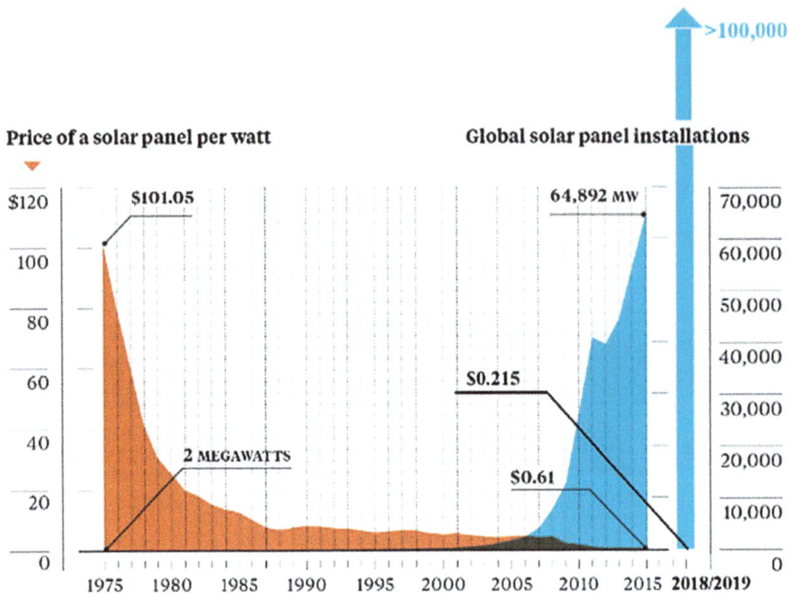

FIGURE B-2 Decline in PV panel prices from 1975 to 2019 as the global solar PV installations increased.
NOTE: Graph adapted from CleanTechnica83 with data from Bloomberg, Earth Policy Institute, GTM, and Energy Trend.
SOURCE: L. Grafman and J.M. Pearce, 2021, *To Catch the Sun*, Arcata, CA: Humboldt State University Press, https://digitalcommons.humboldt.edu/textbooks/8. CC BY-SA 4.0.

TABLE B-1 Summary of State Net Metering Reform Decisions

State	Year of Decision	Initiated by	Applicability	Netting Interval	Excess Generation Credit Rate	Additional Fees
Arkansas*	2020	Act 827 of 2015, Act 464 of 2019	All	Annual (Net Metering)	Retail rate	Grid Access Fee for larger customers (set at zero initially); *struck down by court in 2022*
Arizona	2016; Implemented in utility rate cases	Utility proposal (2013) and ACC (2014)	APS: Residential solar TEP/UNS: Residential and small commercial solar or wind	Instantaneous (Net Billing)	Phasing down to avoided cost rate by 10% per year	APS: Grid Access Fee ($0.93/kW DG capacity *(removed by ACC in 2021)*, *OR* On-Peak Demand Charge. TEP/UNS: Monthly Meter Charge of $2.23 (residential) and $0.90 (non-residential)
California*	2016	A.B. 327 of 2013	All	Annual by TOU period (Net Metering)	Retail TOU rates	Per-kWh non-bypassable charges, upfront interconnection fee
Connecticut	2021	S.B. 9 of 2018, H.B. 5002 of 2019	All	Annual (Net Metering)	Retail rate	Per-kWh non-bypassable charges to be implemented
			All	Buy-All/Sell-All	Sell-all rate determined annually (2022: 29.43 cents/kWh for residential, 20.097 cents/kWh for non-residential); low-income adder of 2.5 cents/kWh and distressed municipality adder of 1.25 cents/kWh	None

continued

TABLE B-1 Continued

State	Year of Decision	Initiated by	Applicability	Netting Interval	Excess Generation Credit Rate	Additional Fees
Hawaii*	2015 (CGS) (*closed*)	HI PUC (2014)	All	Instantaneous (Net Billing)	Avoided cost rate	None
	2015 (CSS)	HI PUC (2014)	All	None	N/A (self-consumption only)	None
	2017 (CGS+)	HI PUC (2016)	All	Instantaneous (Net Billing)	Avoided cost rate (12-mo. avg. on-peak avoided cost)	None
	2017 (Smart Export)	HI PUC (2016)	All paired with battery storage	Instantaneous (Net Billing)	For 4 pm–9 am only: average marginal cost for this time period	None
Iowa	To be implemented after July 1, 2027	S.F. 583 of 2019	All	15-Minute (Net Billing)	Value of solar rate	None
Illinois	To be implemented January 1, 2025	S.B. 2814 of 2016, S.B. 2408 of 2021	Customers not on hourly pricing	Instantaneous (Net Billing)	Energy, capacity, transmission, and purchased energy adjustment components of retail rates (plus upfront rebate)	None
Indiana*	2021/2022	S.B. 309 of 2017	All	Instantaneous (Net Billing); *2022 court decision changed to monthly for Vectren*	1.25 × avoided cost rate	None

Kansas	2018; *struck down by court in 2020 (closed)*	Utility proposal (2018) preceded by KCC investigation (2016)	Residential	Monthly (Net Metering) *(Note residential DG rate features reduced retail rate)*	Reduced retail rate (residential DG schedule includes lower rate of 4.584 cents/kWh)	NCP Demand Charge: $3.00/kW (Oct.–May), $9.00/kW (June–Sept.) for 60-min. demand
Kentucky	2021	S.B. 100 of 2019	All	Monthly (Net Metering)	KY Power: 9.746 cents/kWh (residential), 9.657 cents/kWh (non-residential) KY Utilities: 7.366 cents/kWh LG&E: 6.924 cents/kWh	None
Louisiana	2019	LA PSC (2015) – cap triggered	All	Instantaneous (Net Billing)	Avoided cost rate	None
Massachu-setts	2018; *Fees overridden by legislation in 2018*	Acts of 2016 Chapter 75, Acts of 2018 Chapter 227	All	Monthly (Net Metering)	≤60 kW: Retail rate Other: 60% of retail rate	Eversource: Fixed charge and demand charge (amount varies by customer rate)
	2018	Acts of 2016 Chapter 75	Solar only	Monthly (Net Metering)	*Alternative option to net metering:* Basic service rate	None

continued

TABLE B-1 Continued

State	Year of Decision	Initiated by	Applicability	Netting Interval	Excess Generation Credit Rate	Additional Fees
Maine*	2017 *(closed)*	ME PUC (2016) – cap triggered	All	Buy-All/Sell-All	Reducing production able to be netted against T&D charges by 10% per year	None
	2019	L.D. 91 of 2019	All	Annual (Net Metering)	Granted to utility after 12 months	None
			Commercial and industrial	Buy-All/Sell-All	Rate established annually by PUC (based on standard offer service rate plus 75% T&D rate)	None
Michigan*	2018; Implemented in utility rate cases	Act 341 of 2016	All	Instantaneous (Net Billing)	Power supply rate (excluding transmission component for Consumers Energy, DTE)	None
Mississippi	2015 *(MS's first DG tariff, not a "successor")*	MS PSC (2011)	All	Instantaneous (Net Billing)	Avoided cost rate plus non-quantifiable expected benefits adder (2.5 cents/kWh) and low-income adder (2 cents/kWh)	None
New Hampshire*	2016	H.B. 1116 of 2016	All	Monthly (Net Metering)	≤100 kW: 100% of energy and transmission charges plus 25% of distribution charge >100 kW: Energy charge	None

Nevada	2015 *(closed)*	S.B. 374 of 2015	All	Hourly (Net Billing)	Avoided cost rate	None
	2017	A.B. 405 of 2017	All	Monthly (Net Metering)	75% of retail rate	None
New York*	2017 (VDER)	NY PSC (2015)	All; required for large commercial and community DG	Hourly (Net Billing)	Value of DER rate (includes LMP and value for capacity, environment, demand reduction, locational system relief)	50% Customer Benefit Contribution (CBC) for residential, 70% CBC for small commercial opting in to VDER; others: none
	2020 (Mass Market)		Residential and small commercial	Annual (Net Metering)	Retail rate	CBC: $0.69–$1.09/kW DG capacity
South Carolina	2021 (Dominion)	H.B. 3659 of 2019	All	Monthly by TOU period (Net Metering)	Avoided cost rate	Residential Minimum Bill: $13.50
	2021 (Duke Energy)		Residential	Monthly by TOU period (Net Metering)	Avoided cost rate	Minimum Bill: $30.00 Non-Bypassable Charge: $0.42/kW DG capacity (DEC) or $0.49/kW DG capacity (DEP) Grid Access Fee: $5.86/kW (DEC) or $3.95/kW (DEP) of DG capacity >15 kW
Texas	2017 (El Paso Electric)	Utility proposal (2017)	Residential	Monthly (Net Metering)	Avoided cost rate	Minimum Bill: $30.00 (standard), $26.50 (TOU) *OR* NCP Demand Charge: $3.16/kW for 60-min. demand

continued

TABLE B-1 Continued

State	Year of Decision	Initiated by	Applicability	Netting Interval	Excess Generation Credit Rate	Additional Fees
Utah	2017	Utility proposal (2014) and UT PSC (2014)	All	Instantaneous (Net Billing)	Transitional credit: 3.4–9.2 cents/kWh, depending on customer type	None
	2020		All	Instantaneous (Net Billing)	Rate including value for avoided energy and avoided generation, transmission, and distribution capacity June–Sept.: 5.160 cents/kWh Oct.–May: 4.462 cents/kWh	None
Vermont	2017	H.B. 702 of 2014	All	Monthly (Net Metering)	Blended residential retail rate; credit adjusters based on REC ownership; system size, and siting applied to gross production	None

NOTE: Asterisks denote jurisdictions that are currently considering or have announced plans to consider additional net metering reforms.
SOURCE: North Carolina Clean Energy Technology Center, 2021 and 2022, "The 50 States of Solar Reports," https://nccleantech.ncsu.edu/the-50-states-reports.

TABLE B-2 Notable Public Power Net Metering Reform Decisions

State	Utility	Year of Decision	Netting Interval	Excess Generation Credit Rate	Additional Fees
Arizona	Salt River Project	2014			
California	Sacramento Municipal Utility District	2021	Instantaneous	7.4 cents/kWh	None
Florida	JEA	2018	Instantaneous	Fuel rate (varies monthly; May 2022: 4.025 cents/kWh)	None
Florida	Lakeland Electric	2015	Monthly (*Note solar price plan features reduced retail rate*)	Reduced retail rate (solar price plan includes lower rate of 2.427 cents/kWh)	Demand Charge: $5.27/kW during peak hours
Texas	Austin Energy	2012	Buy-All/Sell-All	Value of solar rate (residential/commercial non-demand: 9.7 cents/kWh; commercial demand (solar <1 MW): 6.7 cents/kWh; commercial demand (solar ≤1 MW: 4.7 cents/kWh)	None
Texas	Pedernales Electric Cooperative	2021	Instantaneous	Value of solar rate (5.377 cents/kWh)	None

SOURCE: Data from North Carolina Clean Energy Technology Center, 2021 and 2022, "The 50 States of Solar Reports," https://nccleantech.ncsu.edu/the-50-states-reports.

TABLE B-3 State and Investor-Owned Utility Additional Fees for DG Customers

State	Utility	Year of Decision	Type	Amount
Alabama	Alabama Power		DG Capacity Charge (for Non-Demand Rate Customers)	$5.41/kW
Arkansas	IOUs, Co-ops	2020; *struck down by Court in 2022*	DG Capacity Charge	$0.00
Arizona	Arizona Public Service	2017; *removed by ACC in 2021*	DG Capacity Charge (for Non-Demand Rate Customers)	$0.93/kW
Arizona	Tucson Electric Power, UNS Electric	2017	Fixed Charge	$2.23 (residential) or $0.90 (non-residential)
Kansas	Evergy	2018; *struck down by Court in 2020*	Demand Charge	$3.00/kW (Oct.–May), $9.00/kW (June–Sept.) for 60-min. NCP demand
Massachusetts	Eversource	2018; *overridden by legislation in 2018*	Demand Charge, Fixed Charge	
New Mexico	Xcel Energy	*Removed by NM PRC in 2018*	Production Charge	$0.0367/kWh
New York	IOUs	2020	DG Capacity Charge	$0.69–$1.09/kW
South Carolina	Dominion Energy South Carolina	2021	Minimum Bill	$13.50

State	Utility	Year of Decision	Type	Amount
South Carolina	Duke Energy Carolinas (DEC), Duke Energy Progress (DEP)	2021	DG Capacity Charges, Minimum Bill	Minimum Bill: $30.00 Non-Bypassable Charge: $0.42/kW DG capacity (DEC) or $0.49/kW DG capacity (DEP) Grid Access Fee: $5.86/kW (DEC) or $3.95/kW (DEP) of DG capacity >15 kW
Texas	El Paso Electric*	2017	Minimum Bill *or* Demand Charge	Minimum Bill: $30.00 (standard), $26.50 (TOU) *or* Demand Charge: $3.16/kW for 60-min. NCP demand

NOTE: * El Paso Electric is a municipal utility, regulated by a municipal board.
SOURCE: North Carolina Clean Energy Technology Center, 2021 and 2022, "The 50 States of Solar Reports," https://nccleantech.ncsu.edu/the-50-states-reports.

TABLE B-4 Types of Benefits and Costs Included in State Value of Solar Studies (by Year)

		Costs		Benefits									
Year	Study	Integration Cost	Admin. Cost	Avoided Energy	Avoided Gen. Capacity	Avoided Transmission	Avoided Distribution	System/Line Losses	Ancillary Services	Risk/Price Hedging	Market Price Suppression	Env. Benefits	Other
2006	Austin Energy (CPR)			■	■	■	■	■				■	
2009	Arizona Public Service (R.W. Beck)			■	■	■	■	■					
2012	Michigan (NREL)			■	■	■	■	■		■		■	■
2012	New Jersey/Pennsylvania (CPR)	■		■	■	■	■	■		■	■	■	■
2013	CPS Energy			■	■	■	■	■		■			
2013	Arizona Public Service (SAIC)			■	■	■	■	■					
2013	Xcel Energy – CO (CPR)	■		■	■	■	■	■				■	
2013	Arizona Public Service (Crossborder)	■		■	■	■	■	■	■	■		■	
2013	North Carolina (Crossborder)	■		■	■	■	■	■		■		■	
2013	Austin Energy (CPR)			■	■	■	■	■				■	
2014	Utah (CPR)			■	■	■	■	■		■		■	
2014	Xcel Energy – MN (CPR)	■		■	■	■	■	■		■		■	
2014	Nevada (E3)	■	■	■	■	■	■	■	■			■	
2014	Mississippi (Synapse)		■	■	■	■	■	■		■		■	
2014	Vermont (Public Service Dept.)		■	■	■	■	■	■			■	■	
2015	Maine (CPR)	■		■	■	■	■	■	■	■	■	■	
2015	Massachusetts (Acadia Center)			■	■	■	■	■	■		■	■	
2015	Louisiana (Acadian Consulting)	■	■	■	■	■	■						■
2015	Tennessee Valley Authority (EPRI)			■	■	■	■	■				■	
2015	South Carolina (E3)	■	■	■	■	■	■	■	■			■	
2016	Arizona Public Service (Crossborder)	■		■	■	■	■	■	■	■	■	■	■
2016	Nevada (SolarCity)	■	■	■	■	■	■	■	■			■	
2016	Nevada (E3)	■	■	■	■	■	■	■	■			■	
2017	Georgia Power (Georgia Power)	■	■	■	■	■		■				■	
2017	District of Columbia (Synapse)	■	■	■	■	■	■	■		■	■	■	
2017	Oregon (PUC)	■	■	■	■	■	■	■	■	■	■	■	■
2017	Entergy Arkansas (Crossborder)	■		■	■	■	■	■		■	■	■	■
2018	NorthWestern Energy – MT (Navigant)	■	■	■	■	■	■	■				■	
2018	Maryland (Daymark)	■		■	■	■	■	■		■	■	■	■
2019	Mississippi (Acadian Consulting)			■	■	■	■	■	■				
2020	Connecticut (PURA)			■	■	■	■	■	■		■	■	■
2021	Pedernales Electric Cooperative			■	■	■		■	■				

SOURCE: North Carolina Clean Energy Technology Center, 2022b, "The 50 States of Solar: 2021 Policy Review and Q4 2021 Quarterly Report," January 2022, https://nccleantech.ncsu.edu/the-50-states-reports.

Committee Meetings

Meeting 1
March 10, 2022
Remote via Zoom
Department of Energy's (DOE's) Office of Electricity: About the study
Christopher Irwin, Program Analyst, Office of Electricity Delivery and Energy Reliability
Perspective of DOE's Office of Energy Efficiency & Renewable Energy
Kevin Lynn, Director of Grid Integration, Office of Energy Efficiency & Renewable Energy
Michele Boyd, Program Manager, Strategic Analysis and Institutional Support team in the Solar Energy Technologies Office, Office of Energy Efficiency & Renewable Energy

Meeting 2
March 15–16, 2022
Remote via Zoom
No open session presentations were held during this meeting.

Meeting 3
April 5, 2022
Remote via Zoom
No open session presentations were held during this meeting.

Meeting 4
April 27–28, 2022
Remote via Zoom
Grid Defection
Jon Creyts, Chief Program Officer, RMI
Mark Dyson, Managing Director, Carbon-Free Electricity Program, RMI
Rachel Gold, Principal, Carbon-Free Electricity Program, RMI
DOE Sponsors
Christopher Irwin, Program Analyst, Office of Electricity Delivery and Energy Reliability
Kevin Lynn, Director of Grid Integration, Office of Energy Efficiency & Renewable Energy

Michele Boyd, Program Manager, Strategic Analysis and Institutional Support team in the Solar Energy Technologies Office, Office of Energy Efficiency & Renewable Energy
Net Metering, Duke Energy & Vote Solar Agreement
Sachu Constantine, Interim Executive Director, Vote Solar
Lon Huber, Vice President, Rate Design and Strategic Solutions, Duke Energy Solutions
Thad Culley, Senior Manager, Public Policy, Sunrun
Rate Design and Net Metering
Scott Burger, Senior Manager, Analytics, Form Energy
Mark LeBel, Associate, Regulatory Assistance Project
Rooftop and Community Solar Industry Perspective
Stephen Campbell, Senior Manager, Policy and Business Development, GRID Alternatives
Sean Gallagher, Vice President of State & Regulatory Affairs, Solar Energy Industries Association
Anne Hoskins, Chief Policy Officer, Sunrun
Mike Judge, Vice President for Strategic Initiatives, Coalition for Community Solar Access

Meeting 5
June 2 and 7, 2022
Remote via Zoom
Hawaii Experience and Perspective on Net Metering
Colton Ching, Hawaiian Electric
Legal Perspective on 3rd Party Electricity Sales
Lee Peterson, CohnReznick

Meeting 6
August 8–9, 2022
National Academies Keck Center
Washington, DC
Regulators Perspective
Marissa Gillett, Connecticut's Public Utilities Regulatory Authority
Daniel Scripps, Michigan Public Service Commission
Value of Solar, Consumer Perspectives on NEM
Karl Rábago, Rábago Energy LLC
Consumer Advocate Perspective
Matthew Freedman, The Utility Reform Network

Meeting 7
September 8 and 29, 2022
Remote via Zoom
Equity Considerations and Net Metering
Roger Lin, Center for Biological Diversity
Michelle Moore, Groundswell
Kim Wolske, The University of Chicago
Gabriel Chan, University of Minnesota

Meeting 8
October 18–19, 2022
Remote via Zoom
Consumer Behavior and Equity in the Electricity System
Gavin Dillingham, Houston Advanced Research Center
Reuven Sussman, American Council for an Energy-Efficient Economy
Discussion with Edison Electric Institute (EEI) Regarding Comments
Shelby Linton-Keddie, Edison Electric Institute

Meeting 9
November 2, 2022
Remote via Zoom
Rural Electric Cooperative Perspective on NEM
Allison Hamilton, NRECA
Consumer Behavior and Equity in the Electricity System
Benjamin Sovacool, Boston University
Solar Industry Perspective on NEM
Sean Gallagher, SEIA

Meeting 10
November 30 and December 1, 2022
Remote via Zoom
No open session presentations were held during this meeting.